Biblical Body Language

The Figurative Face of Scripture

John M. Shackleford

University Press of America,® Inc.
Lanham · New York · Oxford

University Press of America,® Inc.
4720 Boston Way
Lanham, Maryland 20706

12 Hid's Copse Rd.
Cumnor Hill, Oxford OX2 9JJ

Printed in the United States of America
British Library Cataloging in Publication Information Available

Library of Congress Cataloging-in-Publication Data

Shackleford, John M.
Biblical body language : the figurative face of scripture / John M. Shackleford.
p. cm
Includes bibliographical references.
1. Body, Human—Biblical teaching. I. Title.
BS680.B6 S43 2000 233'.5—dc21 00-064889 CIP

ISBN 0-7618-1852-9 (cloth : alk. paper)

™ The paper used in this publication meets the minimum requirements of American National Standard for Information Sciences—Permanence of Paper for Printed Library Materials, ANSI Z39.48—1984

To Jeanne

**For her patience and support,
and who is always there for me**

CONTENTS

Abbreviations		vii.
Preface		ix.
Introduction		xiii.
Chapter One	**The Body of Flesh and the Spirit of God**	1
Chapter Two	**The Biblical (Hebraic) Heart**	15
Chapter Three	**The Head of the Body**	29
Chapter Four	**The Eye**	41
Chapter Five	**The Ear**	53
Chapter Six	**The Mouth**	67
Chapter Seven	**Hand and Arm**	85
Chapter Eight	**The Lower Extremity**	97
Chapter Nine	**Flesh and Blood, Bone and Sinew**	109
Chapter Ten	**Inward Parts, Outward Beauty**	119
Chapter Eleven	**The Resurrected Body**	130
Chapter Twelve	**A Partnership of Science and Religion**	145
Bibliography		153

ABBREVIATIONS

Textual references to Bible versions will indicate the source of the quotation. The Revised Standard Version (RSV) is the source of all block quotations unless otherwise indicated. The abbreviations below will be used to indicate versions of the Bible other than the Revised Standard Version.

IB	Interlinear Bible
JB	Jerusalem Bible
NAB	New American Bible
NRSV	New Revised Standard Version
RSV	Revised Standard Version

PREFACE

This book is not about rational explanations, even if such were possible, of the human body and its relationship to the divine. The intent, rather, is to deepen the appreciation of this central mystery and to show how it makes sense from a faith perspective. Such an objective, of necessity, involves a study of biblical imagery, especially the body language of our Hebrew forebears; how they understood their experiences with the living God, and how they used bodily imagery to communicate their beliefs to others. What is certain, biblically speaking, is the important role of the body in God's strategy in saving history.

I was motivated to write this book because it represents a view of the Bible that, in some ways, is unique. That is not to say that others have not examined the prophetic voice from the perspective of human weakness and bodily structure. In this work, however, I have traced vaious bodily structures throughout the the books of the Bible, from beginning to end. Through this effort I believe we can arrive at an understanding of biblical interrelationships that have been largely ignored in other Scriptural studies.

Biblical authors were able to overcome many of the limitations of human language through bodily imagery. A body part, of course, can be just that, an anatomical structure. In the Bible, however, the human body is most often referred to symbolically. The feet, hands, eyes, ears, for example, become a springboard for launching a profound way of thinking that goes beyond human vocabularies. Thus, physical blindness is contrasted with spiritual blindness, while ears that do not hear are compared with those open to the word of God. The feet may be exalted because they bring good tidings of the Lord. The hand becomes a symbol of power over one's neighbors, or an instrument of conveying one's blessing on another. The mouth can be used as a weapon, or to spread a message of love. Eating, though physiologically important, becomes a highly symbolic event, from the Passover meal in Egypt to the Christian Eucharist. The Hebraic heart, the most important bodily image in the Bible, is the focal center of God's relationship to humankind.

Our ancestors saw, spoke, listened, and acted in terms of bodily language. For them the five senses united bodily humankind with the world around them, and made it into a tangible universe. Paul's revelation that God's power is made perfect in human weakness (1 Cor 12:9), places the entire history of Israel in a new light. Indeed, the strength, dignity, and great honor of being created in the image of God (Gn 1:27), must be re-examined in the context of human frailties, physical as well as moral. From this perspective we can see human flesh, human weakness, as the central issue in Israel's failure to keep its contract with Yahweh. In the absence of human frailty, there would never have been a Babylonian exile. Indeed, humankind would still be luxuriating in the Garden of Eden.

The approach to writing this book is my own, although I cannot ignore the many influences that have seasoned my thoughts over the years. Certainly my background as an anatomist, including the teaching of medical students for over thirty years, has colored my view of other subjects. It was quite natural, therefore, that I should take an interest in the references to human structure found in the Bible. From the beginning of my theological training I sought ways to use both disciplines (biblical studies and anatomy) in some new and creative way. If I have succeeded in this goal I cannot ignore the many influences of the past, especially my theology instructors at Spring Hill College, Drs.Tim Carmody, John Hafner, James Frush, and George Gilmore. I thank Father Chris Viscardi, the department head at Spring Hill College, and Father William Harmless for their continued support and encouragement. Dr. Donald Berry, from the faculty at Mobile College, also deserves my thanks and gratitude for his help and encouragement.

John M. Shackleford
June 27. 2000

Nihil Obstat
Reverend J. Alex Sherlock
Censor Librorum

Imprimatur
Most Reverend Oscar H. Lipscomb
Archdiocese of Mobile
October 13, 1999

INTRODUCTION

In the course of my career as an anatomist I can look back upon thirty-one years of educating thousands of medical and dental students. During those years, it was my job to stress the importance of the human body in understanding the scientific basis of their respective professions. To this end, hundreds of human cadavers were dissected and thousands of human tissue slides were studied in great detail. My colleagues and I taught "normal" anatomy. To understand the changes brought about by disease and injury, and most of all the pathologic conditions they would encounter in later courses, the students must first understand the normal appearances. But what is a normal cadaver? In a strict sense these human remains would not be gracing our dissecting tables if they were, or had been just prior to death, indeed, normal. It goes without saying, these former individuals had to die of something! Thus, paradoxically one might say, we used dead human tissues to teach students the anatomical structure of living human beings.

I remind the reader of this to make an important point. We often use words to describe things and situations which may be wholly, or at least partially, inaccurate. We do this not because we purposely misrepresent the facts, or out of some misguided desire to mislead the other person or persons. Rather, such things come from our lips because of the serious limitations of human languages. Exactly what is "normal," in any given case, whether it be anatomical, physiological, psychological, sociological, etc., turns out to be how we define the term. My "normal" temperature may be a low grade fever for another individual. This person's "normal" behavior may be construed as pathological for someone else.

Often one can be best understood if his/her words are supplemented with body language. That is, some kind of hand gesture, facial expression, or tone of voice, which conveys emotional content, consciously or otherwise, to the overall expression. To be sure, body language is so important and so pervasive in everyday conversation that we tend to take it for granted. We hardly think about it unless another person makes a point of bringing it to our attention. What is even more important than the gesture itself is the symbolic content of body language.

We are so steeped in symbols because we live and communicate with one another in a largely symbolic world. Perhaps this would not be the case if it were possible for human languages to express our exact thoughts and meanings. For better or worse, and God only knows which, this seems to be impossible for the human species.

A "symbol" may be defined as the thing itself and something more, because it carries with it a meaning that goes beyond the physical object, sound, artistic composition, etc. A wooden cross, pursed lips, a bowed head, or a musical composition, for example, may fit this definition. A symphony consists of groups of musical notes, written in a particular sequence, played at a certain tempo. As it is performed, certain notes and tonal qualities receive more emphasis than others. Fortunately, a good symphony is much more than the sum of its musical notations, just as we, as human beings, are so much more than the sum of our parts and bodily functions. By analogy, the human body is like a symphony. No matter how much we know of the nervous system, or whatever structural and functional elements of the body one might analyze, we are so much more, infinitely more I believe, than any analysis can reveal concerning our human biology.

This conclusion is one of the more important aspects of the religious faith that many of us share. Thus, our humanity, or more correctly, the study of what it means to be human, is at the very core of most theologies. We seek to know God and we have a thirst for immortality precisely because we are human. Consciousness of our bodies and the flesh of which they are composed plays no small role in the human appetite for life after death. Without some level of consciousness of our human frailties, we would not know what it is to desire something more. Had we been crafted as immortal beings from the outset, why would we puzzle and meditate about the possibility of the experience? Thus, the very fact of our humanity, the fleshly dwelling place of the human psyche, forces us to look beyond these infirmities and to contemplate other meaningful potentialities.

Nevertheless, there is a lingering tendency to diminish the

importance of the human body vis-à-vis its role in religious goals and expectations. Here, I am not speaking of some ancient heresy in which the body not only takes a back seat to religious consciousness, but is actually construed as being evil. Such excesses, which inevitably lead to denying the fully human Jesus, are rarely encountered in modern religious thought. And yet, the persistence of body-soul dualism, the popular notions of "out-of-body" experiences, and other such examples, lend themselves to this perception of bodily insignificance. The idea of the body as a husk, a "throw away" container for the soul, seems to have gained popularity in our present age of disposable bottles, cameras, contact lenses, flashlights and needles. This diminished emphasis on the importance of the body is certainly not a purposeful designation, or even a conscious one. Yet, one certainly catches a glimpse of this bias from the orations of various public figures, from podium to pulpit, from "Twilight Zone" to TV evangelist.

These days we hear so much about our sinful nature, our preoccupation with material wealth, gluttony, and sexual excess, that it seems quite natural to view the body from a negative perspective. Indeed, we are exposed to this negativity to such an extent that it completely blocks out, or almost so, the positive nature of humankind in the flesh. When Scripture speaks of these weaknesses it is often the natural response to think of the body as an exaggeration of its most degrading moments. Misinterpretations of the Pauline corpus in this regard, tend to deny a central theme in the Apostle's theology, i.e., that God's power for salvation can best be understood as working, even "made perfect," through human weakness (2 Cor 12:9-10). Paul's "when I am weak, then I am strong" aphorism (2 Cor 12:10), should not be viewed simply as a clever play on words. Rather, it reflects Paul's recognition that God's wisdom manifests itself through human weakness.

What matters here, I believe, is that human flesh is a showcase for our dependence on God's saving power; that the physical body has an important role to play in every human endeavor, from the ordinary, everyday tasks, to the most profound peregrinations of the intellect, from the trivial to the sublime.

All bodily behavior gives meaning to the world in which we live. My hands, feet, eyes, etc., realize my intentions and make them come alive in the tangible universe. Who can ever be sure a thought or idea is real, unless it is brought to light through the pen or the spoken word. Encounters with our neighbors reveal the role one's body plays in communication, through speech and gesture. We become a part of one another's world through the language of the body, by means of voice and intonation, sign and symbol. The bodily gesture is not simply a marginal translation of a thought. Rather, it is my anger, love, joy, happiness, etc., in an incarnate mode of expression. To the extent that we are spiritual beings, our bodies are the incarnation of that spiritual essence, just as surely as Jesus is the incarnate God. Thus, the human body belongs to the material and spiritual orders, not as opposing forces, as some might imagine, but in a unity of purpose that may be partly subject to definition, and most certainly, part mystery.

As I said before, to me the human body is like a symphony. One cannot look upon it as a series of notations and derive any sense of its real charm, or more particularly, its transcendent beauty. We are so use to thinking of the body as a collection of parts, our lives as an agglomeration of emotions and functions, that we lose sight of its unifying significance. It is the whole that comprises and mediates its beauty. It is the recognition of what it means to be authentically human, difficult as this may be, before we can even hear the symphony, much less listen intently so as to capture its beauty. Unfortunately, our world is so steeped in dualistic, body-soul, language, that we cannot escape its use even to relate the shortcomings of this kind of thinking. Dualism forces the body's dialogue with the world to become fragmented, so that the symphony is reduced to chaos and confusion. It becomes a cacophonous rendition of the "sinful" flesh.

Politicians are in the habit of describing the poor as "struggling to keep body and soul together." As an anatomist I am not so ingenuous as to believe that our flesh today will be the "flesh" of immortality. Our bodies are constantly losing cells and tissues to the process of "wear and tear." As long as we live, these dying cells, or most of them, are replaced by new ones. With the exception of certain tissues (the central nervous system, for example), there is a constant turnover in the cellular makeup of the body throughout the life of the individual. This marvelous interplay of cells, tissues and organs, ends at death, and the process of tissue corruption begins. We all understand the physical consequences of death. That is why the flesh of the corpse is buried hurriedly, cremated, or embalmed for a more leisurely period of remembrance prior to burial.

And yet, while we are among the living, our seeing, hearing, touching, olfactory sense, etc., unite bodily humankind with the world through the flesh. As Francis Ryan puts it, "There is an internal coordination and collaboration of all my senses before any human consciousness thematically knows about it (The Body as Symbol, p. 35). One might further conclude, as does Ryan, that just as the "manifold experiences of sensate life" are united in one body, so are we also united through sensory experience with the flesh of the world.

The believer, while recognizing that death brings on corruption of the flesh, must also deal with the paradox of bodily unity in the living being. Thus, faith is grounded in a milieu of rational inaccessibility, where human flesh and the divine exist together in the one body. We cannot escape the mystery that surrounds this phenomenon, nor can the anatomist find it and dissect it in the human cadaver. More importantly, that is why the fleshly appearance of God's son is a central mystery of the Christian faith.

What is the role of the body in relation to God's strategy for humankind? How are ideas of our humanity, and its importance to salvation, reevaluated and reinterpreted in the process of a growing religious consciousness? In this kind of study, I believe it is possible to glimpse a facet of the mystery surrounding human flesh and human frailty, as these relate to the idea of immortality. On these hopeful terms, and to that end, this work will proceed to examine the body, and bodily imagery, as it occurs throughout Scripture. At the very least, it is my hope that the reader should become familiar with how our ancestors saw, spoke, listened, and acted in terms of bodily language. In this way the reader will come to have a greater appreciation of who we are as individuals. After all, modern populations did not arise out of a vacuum. We still owe a great deal to our ancestors, even in our unconscious moments, because our thinking is built on their way of thinking. It may be a natural tendency to think of everything in the ancient past as something we left behind forever, as though our ancestors' approach to life holds nothing of value for us today. The other side of the coin, however, tells us that we cannot expect to build on the experiences of the past unless we, in some measure, succeed in understanding them.

CHAPTER ONE

The Body of Flesh and the Spirit of God

> *Yet among the mature we do impart wisdom, although it is not a wisdom of this age or of the rulers of this age, who are doomed to pass away. But we impart a secret and hidden wisdom of God, which God decreed before the ages for our glorification*
>
> (1 Cor 2:6-7)

There is a game at a certain age that seems almost universal in normal children. I played it myself, as did my children, who also went through this phase. It is called building a "fort." A fort can be as simple as a card table with a blanket draped over it, or it may be a more elaborate tree house. The boys usually opt for the tree house, while the girls are usually satisfied with a walled-off corner of the playroom, or the card table fort. I am not a child psychologist, but I have come to some conclusions about fort-building from my own experience.

The fort is a temporary escape from the watchful eyes of Mom and Dad, or it is simply a little piece of territory we can claim as our own, for a short period of time, at least. As such, it adds another shell to our skin, another barrier. When we are young and feeling vulnerable, this little mechanism of escape, the extra protection of the fort walls, however flimsy, does something for our growing personality. It's not the same as our mother's arms, of course, but it is a good, if temporary, substitute. When they are really hurt, children quickly leave the "protection" of the fort and head for the arms of Mother or Daddy.

As we grow older, we construct a different kind of fort, one which does not allow others to enter (unless we want them to). As an adult, our fort is our persona. What the persona does for us is not too different from what the child's fort does. It adds an external barrier, a wall, that tells the public something we would like them to believe about us. Who we actually are is usually not the same image as our persona projects to the public. I would imagine there are some individuals so protective of their private lives that they come to believe their own personas. They don't want to be the persons they really are! They want to be their personas. In that case they never let anyone else inside their fort. They have few if any friends, because a friend would naturally come to understand that the facades they project to the outside world, their personas, are not what they are like at all.

Such an affectation should be more difficult for the faithful Christian. As a Christian I believe I have God's own Spirit within me. As Paul says, our bodies are temples of the Holy Spirit (1 Cor 3:16). Can this be true? If so, is there someone else in the fort with us? Indeed, the person who resides in the fort with us is no less than the Spirit of God. We Christians believe we share our bodies with another person, a divine person. Furthermore, this sharing is so intimate, so dynamic, that we are unable to divide our fort-bodies into two separate compartments, one for us and one for the Holy Spirit. Thus, we are a great deal more than mere flesh and blood, and happily so!

It can be said that this divine presence resides in the heart. Doesn't Paul tell us that the love of God has been poured out into our hearts by the Holy Spirit which has been given to us? (Rom 5:5). But what is the heart? We are not talking here of an anatomical heart, the one that pumps blood throughout the body. We are speaking rather of one of the most important metaphors in the Bible (See chapter two). The function of this biblical heart is not to pump blood, but to infuse the Holy Spirit throughout our bodies, such that there is no separation, no separate room for this gift from the divine Giver. Moreover, this metaphorical heart mediates the dynamic union between flesh and Spirit. Doesn't this change our whole idea of who we are? It should and does, as I hope to show in the pages to follow.

Before we explore the various anatomical details, the eyes, ears, mouth, feet, etc., I am devoting this first chapter to a consideration of the body as a whole. This approach should be more productive, I believe, since subsequent discussions of individual parts can be seen in the proper context of the whole human being. The word, "whole" should be construed to include the human essence as well as those aspects of the human corpus which may be anatomically definable. This is an important consideration, as I hope to make clear, since the idea of what constitutes “the whole” human being changes over time.

Life after Death in the Old Testament

It is generally agreed among scholars that there was no clear idea of a life after death in the Old Testament before the second century BC (McKenzie, NJBC, p. 1313). To the Israelite, Sheol was like a vast tomb where the bodies of the dead lay inert. Like other Mesopotamian religions of the time, there was nothing to look forward to in terms of eternal bliss, or justification of wrongs done to them in life. One could argue that certain psalms make reference to life after death (cf. Pss 49 and 73), which express the hope that there is, in death, some advantage of the righteous over the wicked. There are other examples, some more vague than others, that might be considered as expressing this hope. As McKenzie points out, the first clear reference to the hope of resurrection occurs in the book of Daniel (12:2), composed sometime between 164-157 BC or during the persecution of the Jews by Antiochus IV Epiphanes. From that point in history forward into New Testament times, the idea and hope of the resurrection becomes all the more clearly defined, especially as seen in the writings of the apostle Paul (1 Cor 15:12ff).

The question is, what happened over the thousand plus years, from the patriarchal period (Gn 12-50) to the time of Jesus, to account for this change in views on the prospect for life after death. To a large extent, this altered viewpoint reflects a change in attitude toward the human body. Indeed, from the Hebrew regard for the body of flesh (Heb., basar) to the New Testament body (Gr., soma) as the temple of the Holy Spirit (1 Cor 6:19), one arrives at an essentially redefined, reinterpreted view of the human body. The change reevaluates the whole idea of what it is to be human. We move from the earthly flesh of the Old Testament to the divine immanence of the New Testament, from the desolation of an inert existence in Sheol, to the indwelling of God’s Spirit in our bodies. From a most important corollary of this new understanding of the body comes the hope of eternal life.

Basic to Greek thought, which to some extent the apostle Paul adopts to express his views on the body, is the distinction made between the flesh (sarx) and body (soma). The body, thereby, can be contrasted with the matter or substance out of which it is made. This distinction is never made in the Hebrew Bible, says John A.T. Robinson (*The Body*, p. 13). In addition, of the eighty bodily parts mentioned in the Old Testament, many of them can represent the whole (figurative language which, in modern parlance, is termed synecdoche or metonymy). But the whole is never given any distinction beyond that of the flesh (basar), which composes the part, organ, etc., and thus, the living person.

Human Flesh and the Image of God

Having said all this, it would be misleading to conclude that the Hebrew writers made no distinction between human flesh and animal flesh, or between humankind and the living creatures over which humans are given dominion (Gn 1:26). Indeed, this Genesis account informs us that human beings, male and female, are made in God's own image. A proposition as to just what this priestly account means in all its details, is perhaps not needed here. The point to be made is that the first chapter of Genesis elevates humankind to a position denoting godliness. Humankind is like God in the possession of free will, and creativity. It could be said that one important godly aspect that is missing from the human scene, at least from this early point in biblical history, is the concept of immortality.

Thus, the full implications of being "created in the image of God," becomes subject to later scrutiny and reinterpretation. To the faithful Christian, humankind is potentially capable of eternal life, a godly attribute hardly mentioned, perhaps not considered possible, among the ancient Israelites. One might argue that the idea of human immortality was thought of in a certain context by the Hebrew writer, but regarded as a thought too audacious to pass his lips. There may have been an appetite for life after death in the mind of the ancient Israelite, but because of the presumptuous nature of the thought it remained an unrealizable goal. I will return to this topic in Chapter Eleven, in which the nature of the resurrected body will be considered.

Body and Flesh as Microcosm

Arriving at the idea of immortality is not an easy concept to swallow, especially since decay and corruption of the flesh are such obvious consequences of death. Again, the changes we see in our understanding of what it means to be "created in the image of God," involves a change in how we look upon what it means to be human. In the process of this reevaluation, the flesh becomes no less corruptible than it was in ancient times, so that the idea of life after death must be grounded in a new kind of thinking regarding the dimensions of reality. It makes sense, from a faith perspective that the human body is much more that the flesh and blood that gives it material substance. This was not necessarily the same manner in which the ancient Israelite looked upon himself or his bodily parts.

> There are six things that the
> Lord hates,
> seven that are an abomination
> to him:
> Haughty eyes, a lying tongue
> and hands that shed innocent
> blood,
> a heart that devises wicked plans,
> feet that hurry to run to evil,
> a lying witness who testifies
> falsely,
> and one who sows discord in a
> family.
>
> (NRSV, Prv 6:16-19)

In the foregoing there are five of the six things "the Lord hates" which are referred to in terms of bodily parts (eyes, tongue, hands, heart, feet). As such, each bodily structure reminds the individual of a personal world in which he/she lives and through which he/she functions. My eyes see, and my tongue tastes, the food I put to my mouth. The Israelite also used the anatomical structures to interact with his neighbor, such as, "my eyes behold the face of the shepherd boy and my tongue speaks smoothly to his ears." In this sense a part of the body such as the tongue functions symbolically to denote both selfhood and the social circumstances in which one is immersed. In Psalm 103:1 the Lord is blessed, says the psalmist, "with all that is within me," literally with "my inward parts." What this says to Harold Fisch, in his wonderful book, Poetry With a Purpose, is that the constitutive parts of our physical existence, that is, the very basis or ground of our self-awareness, "serves here to bridge the gap between individual and community—it is somehow common to both."

Among social scientists it is hypothesized that a symbolic relationship exists between society and the physical body as "macrocosm to microcosm" (Neyrey, Paul in Other Words). The body, therefore, can be viewed as a model which represents any bounded system, such that the manner in which the body is perceived relates to how the cosmos is perceived. A key concept, supposedly derived from this relationship, is that bodily control is somehow an expression of social control.

But how might all this relate to the experience of the Israelites? It has been shown (from what has gone before) that Old Testament characters frequently did something, ascribed or relegated a specified action to a bodily part, e.g., my "lips" say this, or his "eyes" behold that, instead of "I" say this, or "he" sees that. From this observation it would seem one could advance the proposition that the Israelite people, through their bodily references, were immersed in their own private world, their microcosmic self. And yet, this view of self seems to lack a certain bodily unity expressed in the personal pronoun, "I."

The New Testament Contrast

> For just as the body is one and has many members, and all the members of the body, though many, are one body, so it is with Christ. For by one Spirit we were all baptized into one body... (1 Cor 12:12-13)

There is a major change in this posture in the New Testament. In the high christology of John's gospel, Jesus identifies himself on several occasions as "I am" (Greek: ego eimi), indicating his divine nature. The most remarkable statement of this occurs when Jesus says to the Jews, "'Your father Abraham rejoiced that he was to see my day; he saw it and was glad.' The Jews then said to him, 'You are not yet fifty years old, and have you seen Abraham?' Jesus said to them, 'Truly, truly, I say to you, before Abraham was, I am'" (Jn 8:56-58; cf. 13:19). In the synoptic gospels, when Jesus came to his disciples walking on the sea, he announced, "It is I (ego eimi), have no fear" (Mk 6:50; Mt 14:27). One can easily read metaphysical philosophy into this expression, not just because Greek was the lingua franca of those days, but because such an expression was unheard of prior to the advent of Hellenism. There can be little doubt that the personal pronoun "I," coupled with the first person, present tense, of the verb "to be," has profound implications as to the identity of Jesus as God's Son, especially, the preexistent nature of that relationship. The only "I am" pronouncement in the Old Testament having comparable weight in its declaration of divine nature is found in Exodus (3:14). Moses, during the theophany at the burning bush, requested God to reveal his name. God's reply to Moses was, "I am who am." Allowing for variations in translation (cf. John Courtney Murray, The Problem of God, p. 10), this utterance is the source of the Hebrew word, Yahweh, the proper personal name for the God of Israel. The

Greek version of the Old Testament (The Septuagint), offers a similar interpretation of, "I am who is [the being]," again the "I am" (ego eimi) representing the all important designation of divine preexistence as the very essence of God. Murray points out, however, that the metaphysical implications of Ex 3:14 did not resonate with the Hebraic ear until after contact with Hellenistic culture (following the conquests of Alexander the Great). Prior to that event the implications of Ex 3:14 were probably more concrete, such as "being historically present to the people in power," or as Murray (p. 10) translates, "I shall be there as who I am shall I be there."

It has been proposed that the Israelite was not aware of himself as an individual in the same sense that we think of ourselves in a modern-day setting. It is the conclusion of Wolff (Anthropology of the Old Testament), that the life of the ancient Israelite was so firmly integrated into his family and the community of his people, that to be set apart or isolated was a very threatening and fearful prospect. It was inconceivably painful, according to Wolff's thinking, for the Israelite to contemplate a personal separation in which he would find himself subsisting as an individual. He would be alone and without recourse to any kind of collective consciousness, especially in his relationship to Yahweh. Contrasting the Christian individual with the Israelite view of self, the Christian is able to relate to his/her immanent God in a way that would seem impossible among the ancient people of God.

Circumcision and the Individual

To illustrate further the idea of microcosmic integration, there is one anatomical feature in particular which best satisfies this understanding, viz., the foreskin of the penis. Removal of the foreskin by circumcision was seen as an act of obedience and, at the same time, constituted a symbol of the covenantal relationship between God and the people (Gn 17:11). Any uncircumcised male "shall be cut off from his people; he has broken my covenant" (Gn 17:14). Notice in the latter verse the warning that if the male is not circumcised (cut off the flesh of his foreskin) God, as a consequence, will cut him off from his people. I do not believe that the language used here is coincidental. The author seems to be emphasizing the relationship of two kinds of cutting, the one, a literal cut (circumcision) and the other, which has more far reaching (societal and theological) consequences. The literal cut, therefore, becomes symbolic of something greater than the physical act, thereby integrating the individual into society in terms of concrete experience. Put in terms of covenantal language, the Israelite male suffers the removal (cutting off) of the penile foreskin to avoid a more painful, metaphorical surgery, which is to be "cut off" from his people and his God. Another type of metaphorical surgery, "circumcision of the heart" (Dt 10:16; cf. Rom 2:29), also bears an important relationship to this "cutting analogy" (see Chapter Two for further discussion).

Individuals in Exile.

Among the Israelites, the limited view of personal individuality begins to change at the time of the Babylonian exile. At that time the people of God had lost temple and monarchy, two of the all important pillars which supported the nation of Israel and their religion. In this difficult period, the oracles of Jeremiah and Ezekiel played an important role in enabling the people to survive the dark days of exile. There is little doubt that these two prophets prepared the way for a new community of Israel, which would one day rise from the desolation the people had experienced in the dark days of exile. The old national-cultic community to which every Israelite citizen belonged had come to an end. If Israel was to survive at all as a recognizable entity, writes John Bright in his book, Jeremiah, "this would of necessity be in the form of a community based far more in loyalty and personal commitment of individuals than the old community had ever been." Just such a community did emerge, says Bright, during the exile and after. The stress of Jeremiah and Ezekiel on the inward and personal nature of the individual's relationship to God "prepared the way for its formation."

This new emphasis on personal responsibility took the onus and fear of wrongdoing from the backs of subsequent generations. Chapter 18 in Ezekiel is entirely devoted to this teaching.

> The word of the LORD came to me again: "What do you mean by repeating this proverb concerning the land of Israel, "The fathers have eaten sour grapes, and the children's teeth are set on edge?" As I live says the Lord GOD, this proverb shall no more be used by you in Israel. Behold, all souls are mine; the soul of the father as well as the soul of the son is mine: the soul that sins shall die (Ez 18:1-4).

The point is further illustrated by examples of father and son. If the father is just and the son is evil, only the son will be held to account (18:5-13). Furthermore, if the unjust son himself begets a son who, seeing the sins of his father, lives a just life, this one shall not die for the sins of his father, "but shall surely live" (18:14-17). Thus only the one who sins shall die, the "righteousness of the righteous shall be upon himself, as the wickedness of the wicked shall be upon himself" (18:20). Ezekiel's teaching goes on to proclaim that there is hope even for the wicked man if he turns away from the sins he has committed (18:21).

Yahweh's Covenant Reexamined

In light of Ezekiel's oracle we begin to see a change, not only in the manner in which the individual viewed his/her relationship to Yahweh, but in the reinterpretation of certain covenantal precepts as well. The first commandment, for example, in which God promises to inflict punishment "on the children of those who hate me, down to the third and fourth generation" (Ex 20:5), would seem to be in conflict with Ez:18. At the very least a new understanding of this portion of the first commandment would seem to be in order.

Likewise, Yahweh's promise to David from the lips of the prophet Nathan, is in need of reinterpretation by the time of the Babylonian exile.

> The Lord also reveals to you that he will establish a house for you. And when your time comes and you rest with your ancestors, I will raise up your heir after you, sprung from your loins, and I will make his kingdom firm. It is he who shall build a house for my name. And I will make his royal throne firm forever. I will be a father to him, and he shall be a son to me. And if he does wrong, I will correct him with the rod of men and with human chastisements; but I will not withdraw my favor from him as I withdrew it from your predecessor Saul, whom I removed from my presence. Your house and your kingdom shall endure forever before me; your throne shall stand firm forever. Nathan reported all these words and the entire vision to David.
>
> (*NAB* 2 Sm 7:11-17, my italics)

Some six hundred years later, just before the Babylonian exile, many of the people felt that neither their nation nor their temple could ever be destroyed, principally because God had promised as much to David. They felt secure, even in their sinfulness and their idolatry, because they took comfort in the words of Nathan's prophecy. And yet, circumstances were going from bad to worse in those days. The Babylonians were growing stronger and were pressing hard at the gates of Jerusalem. Enter the prophet Jeremiah, who tells the people in so many words that their assumptions concerning the forbearance of Yahweh are dead wrong.

> Thus says the Lord concerning the king who sits on David's throne, and all the people who remain in this city, your brethren who did not go with you into exile; thus says the Lord of hosts: I am sending against them sword, famine and pestilence. I will make them like rotten figs, too bad to be eaten. I will pursue them with sword, famine and pestilence, and make them an object of horror to all the kingdoms of the earth, of malediction, astonishment, ridicule, and reproach to all the nations among which I will banish them. For they did not listen to my words, says the Lord, though I kept sending them my servants the prophets, only to have them go unheeded, says the Lord. (*NAB*, Jer 29:16-19)

There is no doubt that these words seem to contradict the prophecy of Nathan spoken to David. At the very least the contrast between Nathan's prophecy and that of Jeremiah has to be understood in a new light, particularly the promise that, "Your house and your kingdom shall endure forever before me; your throne shall stand firm forever." Even David seems to have interpreted these words in a literal sense when he says just before his death, "Is not my house firm before God? He has made an eternal covenant with me, set forth in detail and secured" (2 Sm 23:5).

Was Nathan a false prophet? Did God renege on the promise to David? Either proposition could be counted as credible to the casual reader, but a third insight can be construed from the Christian viewpoint. What is central to the covenant here is not the preservation of strict interpretations of Nathan's words. The central issue is God's promise of a faithful relationship with his people. God did not say the people would not suffer, especially if they were disobedient. The unconditional love of God does not mean that the response of the people is irrelevant, that they could break their obligations to the covenant and do whatever they chose, including various forms of idolatry, and not suffer the consequences.

The Christian Understanding.

To Christians, Yahweh's promise regarding kingship and temple speaks of an entirely different dimension of reality. The literal interpretation of Nathan's words doesn't begin to capture the full implications of what is to transpire over the next thousand years. Examining saving history from this point of view, we move from the period of David's kingship to the kingship of Jesus the Christ, from the idea of a physical temple of stone and mortar to the temple of the Holy Spirit, from holocaust offerings to the Christian Eucharist.

One might ask, would this new understanding have come about without suffering? Would Israel have looked at their religious suppositions, reexamined their relationship to Yahweh, in the absence of the Babylonian exile? The fact is, the people of God broke their half of the contract. Had they not disobeyed their obligations to the covenant, one might then ask, would there have been a Babylonian exile? Would the suffering of the people have been necessary in that case?

Knowing the weaknesses of the human flesh, the latter scenario seems highly unlikely. To be sure, God knows of our weaknesses better than we do ourselves. Therefore, the suffering, the exile, followed by more suffering, are all the more likely. It is part and parcel of the process of spiritual growth. To use an old analogy, human suffering is the anvil and God is the blacksmith. Reading the Bible from this point of view is doubly enriching, I believe, because one can trace the forging of our religious consciousness through saving history, following it through humankind's early nomadic wanderings, right down to the Christian era.

The transformation in human consciousness is striking when we take the time to look at it from a developmental perspective. Freedom from slavery in Egypt, a physical thing, becomes a metaphor for freedom from the slavery of sin. The nomadic peregrinations through the Sinai desert are construed by Christians as a symbolic reference to life's journey, in which one resists or falls prey to sinful temptations at every turn. The Israelites' golden calf is now likened to any priority we place before God, such as money, power, sex. Modern idols, although more abstract, are just as powerful as the ancient ones. Finally, the "Promised Land," the final goal of the wandering Israelites, is the most powerful symbol for each one of us struggling to remain faithful to our God. The Christian faithful travel through the desert of human temptation and sin, striving always, in spite of stumbling here and there, to reach the promised land of eternal life, God's kingdom.

All the past experiences were necessary. Otherwise, why is there such a thing as history? How would the people have accepted the notion of a kingdom of God, without experiencing a human king and a nation of Israel? To be sure, there would have been no basis for understanding such an abstract concept.

History Driven by Human Weakness

In many ways the Babylonian exile was a crucial turning point in the history of the people. Losing the temple and the monarchy, as has been said, forced the Israelites to reexamine all of their previous assumptions, especially those based on the law of Moses and the prophecy of Nathan. As we read the prophecies of Jeremiah we learn that there is to be a new covenant, one that will take precedence over the old. God is going to intervene in history in some new and unprecedented way, for the law will be written on the people's hearts (Jer 31:33). With the loss of temple and monarchy, and the suffering of exile, there is also to be a new relationship with Yahweh, one which stresses individual responsibility (Ez 18:4). All and all, these prophecies must have generated in the people an outlook of hopeful expectation unlike any they had ever experienced in their earlier history as a more corporate Israel.

In this transformed sense of identity, the people will not only be partners in a new covenant, but they will be given a new heart, and a new spirit will be put within them in place of the old stony heart (Ez 36:26). It takes no great leap of the imagination to conclude that the prophets, particularly Jeremiah and Ezekiel, saw the individual human being in a new light. The body of flesh and blood had been elevated in status in a new relationship with Yahweh. A new spirit will be put within human flesh, and God's laws will be written on the flesh of the human heart.

Paul's Strength Through Weakness.

It is perhaps unfortunate that we tend to read St. Paul, the premier theologian of the Christian Church, only in the context of his letters. The advantage of applying Paul's revelation concerning God's power made perfect in weakness (1 Cor 12:9), for example, places the entire history of Israel in a new light. From this perspective we can see human flesh, human weakness, as the central issue in Israel's failure to keep its contract with Yahweh. In the absence of human frailty, there would never have been a Babylonian exile. Indeed, humankind would still be luxuriating in the Garden of Eden.

But if that were the case, would we really be human? Once

given the freedom of choice, the ability to choose between good and evil, sin and obedience, love and hate, humankind was bound to exercise those options. The privilege of free will was given to us, as theologians tell us, because without it we could not truly love God. God could have made us into automatons, glorified biochemical machines, programmed to obey always, and there would have been no such thing as human weaknesses. If we were some kind of machine, mechanical failure or disruption of the biochemical mechanisms would have been the order of the day rather than moral decline or failure.

But that is not what God wanted when we were created in God's image, for without the option to hate, humankind could not choose love. Without evil how can there be goodness, freely chosen? How could there be genuine obedience without the option to do otherwise? But God did not create automatons, beings who would follow a preordained path programmed into the psyche, because the Lord wanted us to freely choose to love him and freely fulfill the obligations to the covenant relationship.

Being created free means making mistakes, sometimes choosing evil over good, sinfulness over obedience, and learning from those mistakes. We learn from our mistakes in the long course of human history only if the mistakes of the past are passed on from generation to generation. And that is what the Bible is all about. The Old Testament does not try to cover up human failures. Indeed, it highlights them. That is the lesson of the prophets, then and now, and that is why these great men were not very popular among our ancient forebears. Nobody likes to have his failures thrown into his face.

The Importance of the Human Body

Perhaps I should apologize for bringing up the obvious. Almost everyone knows these things about sin and disobedience. What may not be so obvious, however, is that the human body is the focal center of our coming to know God better. The newly understood status of the individual in the aftermath of the Babylonian exile brought with it a greater appreciation of human dignity, individual human dignity. The Israelites, among all the peoples of the ancient world, began to think of themselves as something more than flesh and blood. Indeed, they were worthy enough in this new context to have God intervene in history in an unprecedented way, to promise them a new covenant, to write his laws upon their hearts. The people were as yet a long way from thinking of the body as the temple of the Holy Spirit (1 Cor 6:19), but they had taken the first step. Without this stepwise growth in the appreciation of their own bodies, their dignity as human beings created in the image of God, they would never have come to appreciate the next step, the concept of the immanent God, the indwelling of the Holy Spirit.

As I hope to show later, this new rise in human consciousness that accompanied the pain and suffering of the exile is also a step toward an appreciation of the possibility of life after death. This would be an unthinkable proposition without a new understanding of the human body as deserving such a glorious transformation. The immensity of God's love for us was just beginning to dawn on the people, and there was a sense of anticipation never before experienced. What does it mean, that God is going to give us a new heart? How is he going to write his laws on our hearts? How can God put a new spirit within us? How are all these things to be accomplished?

CHAPTER TWO

The Biblical (Hebraic) Heart

> *More than that, we rejoice in our sufferings, knowing that suffering produces endurance, and endurance produces character, and character produces hope, and hope does not disappoint us, because God's love has been poured out into our hearts through the Holy Spirit which has been given to us.*
> (Rom 5:3-5)

These powerful words of Paul are more than simply prophetic; they are a formula for God's strategy for salvation. In the Christian faith, it is possible to rejoice in our sufferings, because the faithful know that suffering is part and parcel of saving history. In suffering one learns to endure, and in so doing, one arrives at the conclusion that this suffering has meaning. With meaning in our lives and in the world, our endurance makes us better people; it gives us the strength of character that would not have occurred in the absence of suffering. Through force of character we have the courage to hope in God's love for us. There is no disappointment in this hope because, as Paul says, God's love has been poured out into our hearts through the Holy Spirit.

C. G. Jung concludes (*Man and His Symbols*), from his psychoanalytic point of view, that "no genius has ever sat down with a pen or a brush in his hand and said:

> "now I am going to invent a symbol." No one can take a more or less rational thought, reached as a logical conclusion or by deliberate intent, and then give it "symbolic" form. No matter what fantastic trappings one may put upon an idea of this kind, it will still remain a sign, linked to the conscious thought behind it, not a symbol that hints at something not yet known.

Jung points out that there are many symbols that are collective, that is, stemming from the "collective unconscious," rather than individual in their nature and origin. These are chiefly religious images among which I would place the Hebraic heart. The origin of such collective symbols, says Jung, "is so far buried in the mystery of the past that they seem to have no human source."

What this Jungian point of view says to me is that symbols, such as the biblical heart, go beyond language and the ability to explain them fully by any rational means. They are rather, part and parcel of our human nature, such that any approach to understanding them requires that we be in touch with the very ground of our existence. The ultimate meaning of the biblical heart, then, has a breadth and depth that touches us as a species, and as special creatures of God, created by God in his own image.

In his book, *The Distancing of God*, Bernard Cooke says human experience is "constantly shaped by symbols," especially through the metaphors we use and their connotations, the overtones of people's actions, the rituals we engage in to relate to one another in families or in society, the artifacts that surround us, and even by the images used in advertising. Our human nature, our very being, believes Cooke, has symbolic depths that we still do not fully understand.

In his assessment, Cooke seems to agree with Carl Jung that the origin of symbols (religious ones in particular) is buried deeply in the mystery of the past. It is my conviction that the heart is a major biblical symbol, also buried deeply in the prehistoric past of humankind. As such it continues to bubble to the surface of the conscious mind, crying out to be recognized.

It does not take a rocket scientist or a professor of anatomy to recognize at least two meanings for the word, heart. First there is the organic heart, the one that pumps blood throughout the body. Second there is the metaphorical heart, that which is the focal center of our human essence. Indeed, the metaphorical heart is that quality that dignifies us as authentic human beings, created in the image of God. As contrasted with the organic heart, the biblical, or Hebraic heart, is not exclusively matter or spirit. It is somehow both. I should like to explain how this sophisticated notion developed out of the inspired thinking of the Old Testament writers.

The Biblical Heart from Myth to Metaphor

To refer to something in the past as a "myth" does not imply anything derogatory. Indeed, the general view of biblical myths is that they contain truths far more sophisticated and meaningful than their literal reading might convey. The story of Adam and Eve, for example, is said to contain much more in terms of "mythic truth" than if we were to believe that Eve was actually tempted by a talking serpent. By the same token, the biblical heart is expressive of much more truth than if we take it as an ignorant misunderstanding of the organic structure, the one that pumps blood. In this section, therefore, an attempt will be made to explain how the heart of the mythic world of the ancient Israelite became the metaphorical heart of today. This is not to say that the biblical heart, itself, is to be taken as myth. Rather, the term is mythic in that it derives from the mythic world of the Bible. As I shall explain later on, the heart is a primordial symbol which is defined in terms of the creation myth of Genesis 1:26, i.e., that humankind is created in God's image and likeness.

The objective here is to recapture some of the truths hidden in the symbolism of the biblical heart. The heart was, after all, the agent through which the ancient Israelites communicated with God. As such, the heart was the focal center of their existence in relationship to God. If one could conceive of a device that had one end related to the material universe and the other end extending into in the divine dimension, through which God could communicate directly with us, and vice-versa, such a device would partially explain how the Hebraic heart works. The heart is therefore analogous to a wormhole in space, in that the heart, instead of connecting different galaxies, or segments of the universe, establishes a link between the material universe with the divine universe.

In spite of the accumulated scientific knowledge concerning the heart (even up to the present day), many of us daily use the name of this organ in a figurative sense. Perhaps this is because, as Rahner and Vorgrimler point out in their *Dictionary of Theology,* the "heart" is a basic concept of "primitive anthropology," and has to do with the center of "the personal spirit's self-control." The heart as a physical organ, therefore, remains for us a "primordial symbol." It probably took no great feat of the imagination, even for the earliest of humans, to associate the beat of the heart with life itself. We, as twentieth century members of the same race of beings, can easily imagine how our present day metaphorical sense of the word, "heart," may have begun there, with what was to the earliest humans a reality and relationship they were proud to comprehend.

In view of the above, the human heart of ancient times, up to and including the Hebrew canon, is less figurative (in its literary applications) than is possible in the present age. When we say, for example, that we love God "with our whole heart," having in mind our present day knowledge of cardiac physiology, the expression would be almost totally figurative (for many, at least). To the ancient Israelite, however, the expression would seem to include a level of non-metaphorical reality that is perhaps difficult for us to appreciate. For one thing they would not have even the rudiments of cardiac physiology to draw upon. The ancient Israelite would have no frame of reference for comparison, as we do, between the organic structure and the "heart" by means of which one expresses a love for God.

In one sense our modern perspective is a good one, since the heart image is no longer clouded by the ancient misconceptions as they relate to the organic heart. In another sense, because of our separation in time from the ancient mindset, we are faced with difficulties in fathoming the original implications of the biblical heart. Here the mythological truths are so fundamental and such an integral part of the human equation that the best we can hope for is a rough approximation. Consider the property of human volition, one of the many aspects of the Hebrew heart, which today many would consider to be more properly associated with the nervous system. But "free will," as it is commonly regarded in the twentieth century, is just as much a mystery to us as it was to the ancient Hebrew writer.

Volumes have been written on the structure and function of the human nervous system, but nowhere in the complex tangle of neurons, ascending and descending fiber tracts, commissures, reticular formations, nuclei, ganglia, etc., has anyone uncovered a function or anatomical location that can be ascribed to "free will." We do indeed have what is called voluntary motor activity with which we respond to various conditioning and stimuli, but these reactions (although they may seem to be freely chosen) do not constitute the kind of freedom that faith tells us only God can bring to the human clay (cf. 2 Cor 4:7). To be sure, Scripture teaches that we are in possession of true freedom only when we rise above our fundamental biochemistry through this extraordinary gift of God. The gift of free will is part and parcel of the image we are now calling the "biblical heart."

Brevard Childs (*Old Testament Theology in a Canonical*

Context) views "all parts of the body" as functioning in the Old Testament in a "metaphorical-like manner" to describe in terms of Hebrew reality "different aspects of total life as a human being." In the case of the Hebraic heart, I am saying that we have appropriated it as metaphor in our present century, and in the process we have extracted it from the mythic world of the biblical people, happily to our great benefit. What was once a major element in their mythic world has become for us a major metaphor. It is a metaphor with a difference, true enough, for we usually associate metaphor with a literal term which, in turn, is identified with a figurative one. When a person says, for example, "he speaks out of both sides of his mouth," the anatomical part, mouth, does not lose its functional character, even within the context of the metaphor. But if, by contrast, we exclaim, "I know this in my heart," the expression has nothing whatever to do with the organic heart. This is one reason, perhaps, that the heart as metaphor might be regarded in a somewhat different light than are other bodily parts which may also be construed as metaphor.

The Hebraic heart, therefore, will always remain at least partly refractory to attempts at demythologization. As should become increasingly evident, it is not a concept that one can explain by asking the right questions. Indeed, the biblical heart is not a concept at all. It is, rather, a primordial symbol which relates to the most basic notions of what it means to be human.

The Heart as Primordial Symbol

In biblical myths the stories almost always have embedded in them some important truth. The Adamic myth (Gn 2), for example, illustrates a case of disobedience against God, or, as it is commonly referred to after Augustine, *original sin.* Indeed, sin, as an early symbol, is defined by the Adamic myth. The interesting thing about a symbol such as sin is that it must of necessity precede in time the myth that defines it, which, in this case, is the Adamic myth.

Some years ago, when I was teaching microscopic anatomy to medical students, one of the subjects covered was the human ovary. There are present in the ovary, beginning with the fifth or sixth month of fetal development, what are called "primordial follicles." These follicles contain a primary oocyte, which can develop into mature ovum, capable of being fertilized once it is released from the ovary. Of course, this cannot happen until the fetus reaches the age of a sexually mature adult. Just thinking of this phenomenon boggles the mind; a female fetus, in the mother's womb just six months, already contains in its ovary the potential for a whole new generation of human beings.

Each of these primordial follicles, when mature, is the first in time to be recognized as the female contribution to a future human person. That, of course, is what the word "primordial" means. It is the first in time, or the earliest recognizable form of something, the original in any developmental sequence. Therefore, when I state my opinion that the heart is a primordial symbol, I mean that it is the earliest symbol one can associate with the human person, even earlier than that of sin. It is the primordial follicle of human symbols.

Norman Perrin in his book *Jesus and the Language of the Kingdom*, discusses primordial symbols at some length. One of the interesting insights to come out of this work is that primordial symbols must be interpreted by myth. Citing a book by Ricoeur, The Symbolism of Evil, he states that the symbol, in this case sin, is prior to the myth (the Adamic myth) which interprets it. What this means is that the biblical myth of Adam and Eve tells a story that, perhaps among other things, contains a truth too complex and abstract to the Yahwist author to express except through the imagery of the story. The symbol interpreted by this myth is sin, in this case, the original or primordial sin of disobedience to God.

In connection with the heart, I should like to examine the other creation story found in the first chapter of Genesis.

> Then God said, "Let us make humankind in our image, according to our likeness; and let them have dominion over the fish of the sea, and over the birds of the air. . ." So God created humankind in his image, in the image of God he created them; male and female he created them
> (*NRSV*, Gn 1:26-27).

This Priestly story of creation, as I hope to make abundantly clear, is the myth that interprets the primordial symbol of the human heart. It is a symbol even earlier, more primordial, than that of sin. This conclusion, I believe, is unavoidable, because one must first recognize a focal point of existence out of which a consciousness of sin develops. This consciousness and focal point of human existence, as I am attempting to show, is and must be the human heart. It is to this focal point that God turns in the act of writing salvation history (cf. Jer 31:33). It is the human heart that God wills to harden (Rom 9:18; Is 6:10) or chooses to give his Spirit into (cf. Jer 36:26; Rom 5:5; 2 Cor 3:3; Gal 4:6). The heart as symbol, therefore, must be the most primordial of all symbols because it has to be present for humanity to function as such. Humanity, as created in the image of God, cannot exist without it, much less commit sins, because the Hebraic heart defines what it is to be human. Neither sin nor salvation is possible in the absence of the biblical heart, because it is into this "vessel" that one takes in idols (false allegiances), or into which the Holy Spirit enters. It is the only "organ" open to the word of God, whether or not the other organs, such as the eyes and ears, are fully functional.

Now if one were to say that the heart symbol is that which is defined by the first creation story (Gn 1), why continue to call it a "heart?" Why not simply call it the "soul," or by some other, "more appropriate" designation? A further objection might be that the heart term is too confusing, since we have an organ that pumps blood by that same name.

I would be among the first to admit that there is much confusion inherent in the term, "heart," especially since so many use it to express their deepest emotions, their most profound sense of compassion, their sincerest convictions, etc., without really understanding the biblical context of what they are saying. It is possible, of course, to substitute a variety of terms that one can define as meaning the same thing as the biblical heart. But that merely states the problem from another perspective, and if one is going to define another word to mean (and take the place of) "heart," then it becomes incumbent upon that person to know and understand the vast range of imagery that he/she has brought to bear on this new term.

The "human soul," for example, would not be such a bad substitute or synonym for "heart" if we were willing to make the change. The fact is, we aren't. The heart metaphor is so ingrained in our twentieth century thinking, in whatever language we invoke it, that we can hardly do without it. We must therefore acknowledge, it seems, that the biblical heart alludes to a depth of feeling and truth so fundamental, that no other term can compete with, or substitute for it. The weight of evidence certainly suggests this conclusion.

Those who so abundantly sprinkle the term throughout their speeches have probably never heard a definition of the metaphorical (biblical) heart. In one sense, few persons are in need of a definition, because most have heard it used so many times by speakers (and authors) other than themselves. It is a term so ingrained in our psyches that we are not really very sure where it came from, or exactly what it means, except that it sounds good and seems appropriate to a given occasion. But this is the nature of a symbol, particularly a primordial symbol such as the heart. It goes so far back into the dim reaches of the past that it is impossible to determine its exact origins. Yet it is present to us, half conscious, and half buried in what Carl Jung refers to as the "collective unconscious." We use it every day without knowing or understanding its deeper meaning, but in some inestimable way, sensing its import nonetheless.

The Idolatrous Heart

To explain the idolatrous heart we must, for convenience sake, think of the heart as a vessel or container of sorts. As such, the heart abhors a vacuum; it must be filled with something. Luke says it well:

> When an unclean spirit goes out of someone, it roams through the regions searching for rest but, finding none, it says, "I shall return to my home from which I came." But upon returning, it finds it swept clean and put in order. Then it goes and brings back seven other spirits more wicked than itself who move in and dwell there, and the last condition of that person is worse than the first
>
> (*NAB*, Lk 11:24-26).

The point here is that one cannot sweep out the idols from the heart without replacing them, without filling the emptiness with something. We simply cannot have an empty heart. Indeed, biblically speaking, it is an impossibility. The ideal, of course, is to replace the idolatrous thoughts with an awareness and openness to the love of God. The love of God is always present in the heart, Christian theologians tell us, but the idols ("demons") crowd out the ability to be aware of that presence. In this I believe, with Father William Shannon (*Silence on Fire*), that we never have to find God, or his love, because he is always there (in our hearts). The task, rather, is to become aware of that presence through prayer. Ezekiel, after a lengthy warning about false prophets who are leading Israel astray, says to certain elders of Israel,

> "Mortal, these men have taken their idols into their hearts, and placed their iniquity as a stumbling block before them; shall I let myself be consulted by them? Therefore speak to them, and say to them, Thus says the Lord God: Any of those of the house of Israel who take their idols into their

> hearts and place their iniquity as a stumbling block before them, and yet come to the prophet—I the Lord will answer those who come with the multitude of their idols, in order that I may take hold of the hearts of the house of Israel, all of whom are estranged from me through their idols. Therefore say to the house of Israel, Thus says the Lord God: Repent and turn away from your idols; and turn away your faces from all your abominations. . ."
>
> (*NRSV* Ez 14:3-6)

Notice that the idolatrous hearts have been the agents of estrangement from God. They have not removed God from the Israelites' hearts, because nothing can take God away from us. God is with Israel, even in the presence of their idols. The idols, however, are that which occupies the peoples' thoughts, not the Lord their God. Israel is therefore unaware that God is with them, so that God must now make his presence known by "taking hold" of their hearts.

Idols, it seems are the "nuts and bolts" of sin. That is to say, without idols in the heart there can be no sin. This may at first glance appear to be a strange thing to say until one begins to reflect on the nature of sin. One does not go out and begin disobeying God, just for the sake of disobedience. Such disobedience, to be sure, must be prompted by some other allegiance, such as wealth, power, sensual pleasure, or something more powerfully present and desirable to the individual than obedience to the will of God. These other allegiances would constitute the idols of the heart. An Idol, therefore, is not necessarily a golden calf, or some pagan god such as Zeus or Apollo. Idols are rarely identifiable with some pagan symbol, or graven image, especially in the present century. An idol can sometimes take a seemingly innocuous form, such as excessive attention to dieting, physical exercise, etc., in order to have a "perfect body." When anything becomes such an obsession for the ancient Israelite, or for the twentieth century American, it has the potential to become for that individual an idol. As such it can lead to allegiances other than to God, i.e., to that behavior we are labeling as "sin."

The Hardened Heart

The imagery of the Hebrew heart as the focal point of God's relationship to his people pervades the entire Old Testament. One particular form of this image, the "hardened" heart, constitutes a recurring motif throughout the Pentateuch, the book of Psalms, wisdom literature, and the historical and prophetic books. The hardened heart of the Old

Testament portrays the idea of stubbornness and persistent obduracy in the face of God's will, although it would be a mistake to limit the image to such a narrow interpretation. The hardened heart, may be thought of as a consequence of consistent and unrepentant idolatry.

Other heart images in the Old Testament are very similar if not identical to that of the hardened heart, e.g., the "uncircumcised" heart (Lv 26:41; Dt 10:16; Jer 4:4; Ez 44:7), and the "fat" heart (Ps 119:70; Is 6:10). The uncircumcised heart is of particular interest because it is necessary for one to remove (circumcise) the metaphorical foreskin of the heart as a remedy for its "hardness" (Dt 10:16; Jer 4:4). This explains the reasoning of the apostle Paul when he writes, "real circumcision is a matter of the heart, spiritual and not literal" (Rom 2:29).

Gerhard von Rad dedicates a significant amount of space in his book (*The Message of the Prophets*) to the hardness of heart theme. "God erected a terrible barrier against Isaiah and his preaching: he hardened Israel's heart." Many commentators find no great difficulty in interpreting this, says von Rad. They fall back on psychology in their appeal to the "undisputed" fact that persistent rejection of God's word leads to mental dullness such that the ability to hear and understand fades away. To habitually ignore the warnings of God inevitably brings on a kind of spiritual deafness. It seems correct that the psychological interpretation becomes a "general truth of religion," comments von Rad, "which can be constantly confirmed in the broad realm of religious experience." But however correct the facts of this psychological phenomenon may be, von Rad objects strenuously to the interpretation of hard-heartedness in this manner.

The conclusion von Rad reaches, after having emphatically rejected the psychological interpretation, is that hardening of the heart is a "particular mode of Yahweh's historical dealings" with the people of Israel. This means, of course, that one must learn to read whatever is said about hardening of the heart in the context of saving history. In Isaiah's case, the message against which Jerusalem hardened its heart points to a future generation. The message which had fallen on completely deaf ears in Isaiah's day will come to be "fulfilled" in the new generation.

To me, von Rad's position on psychology is somewhat overstated. It seems that from the human perspective we have two choices in attempting to understand the "hardening" process. Either God interacts with his human creatures through their innate composition, i.e., through their hearts (or minds in the case of a psychological interpretation), or he must transform them into a totally new species of automatons. If he chooses to recreate humans as hard-hearted beings; if this is accepted as a possibility, it appears to me that acceptance of such an event would also lead to spiritual and intellectual bankruptcy. "Re-creation" of humanity (or that segment of humanity God wills to control) as hard-hearted beings, would remove human freedom from playing any role in saving history. It is possible, therefore, that von Rad is rejecting the psychological explanation only to the extent that it takes God out of the equation. I therefore believe that for saving history to have any meaning to Christians, one cannot remove either God or human volition from their respective roles, albeit God undoubtedly having the major one in implementing a salvific strategy.

Otherwise, von Rad's interpretation of hard-heartedness certainly rings true if applied to Paul's letter to the Romans. When Paul writes (9:18) that God "hardens the heart of whomever he will," he makes no effort to explain this hardening as something a person brings upon himself. In fact, he relates the concept to his particular historical circumstance, i.e., that God's hardening of Israel is related to bringing the gentiles into the fold of Christ (Rom 11:25). In this viewpoint, Paul recognizes that Israel's lack of belief, i.e., their rejection of Christ, is related to God's overall salvific strategy, which eventually, as Paul believes, will include the Jews as well as Gentiles (Rom 11:26-27).

The Hebrew writer recognized, believed in his heart, that God was present to his people in history. One way the ancient writer chose to express that belief was in the hardened heart imagery. Just how that imagery translates into what God actually does, is in my opinion, beyond human comprehension. It is among the deepest of mysteries. It seems appropriate to postulate that, in whatever manner God chooses to exercise influence in human history, such acts as these are accomplished without destroying human freedom, or at least they are done without the abnegation of that faculty which allows one to say "yes" or "no" to God's love. In the absence of this faculty, history becomes meaningless.

A New Heart and a New Covenant

In view of our weakness of the flesh, idolatry, and hard heartedness, it would indeed be a bleak outlook for the future if this were the whole story of humanity's relationship with God. But in the face of this dismal picture, Paul says that hope does not disappoint us, "because God's love has been poured into our hearts through the Holy Spirit which has been given to us" (Rom 5:5).

"Prophecy and fulfillment," is perhaps a key phrase in getting at the central theme of how the early Christian church reexamined the oracles of the Old Testament prophets. More to the point, how were the New Testament writers justified in applying these prophecies to Jesus? The author of Matthew's gospel, in particular, made frequent use of the formula: *Such and such an event took place in order to fulfill what the prophet said.*

It should be abundantly evident, from the biblical examples that have been presented thus far, that revelation from God comes to the human creature in and through that magnificent faculty we have come to know as the "heart." The heart, however, is not like the intellect in receiving these "words" of inspiration. There is no tape recorder, and there are no verbatim translations possible in divine revelations. As far as anyone can tell, the prophetic voice is not a word-by-word playback of words God has spoken.

In twentieth-century theology the idea of symbolic disclosure in revelation has been widely accepted. Avery Dulles and Karl Rahner are among those who have given great credence to this mode of revelation (Dulles, *The Craft of Theology*). Moreover, these symbols may be in the form of words or images that the prophets must, in turn, interpret for the masses. The presence of symbolic elements at turning points in salvation history, such as contained in the "new covenant" and "new heart" prophecies of Jeremiah and Ezekiel, are good examples.

At some point, however, symbolic disclosure must be laid bare in a form open to human interpretation. The symbolic content, in fact, usually spills over into the written or spoken syntax, in which case one can only guess at the nature of the original revelatory content. To me, this means that even the prophet himself may have been unable to fully understand the precise meaning of his prophetic utterance, especially insofar as the oracle might apply to a circumstance in the distant future. For the most part, the prophets were in the business of interpreting the times in which they lived, and not in predicting the future.

In view of the imprecise nature of the revelatory event, it would be foolish to limit oneself to a one-dimensional interpretation, i.e., the particular interpretation of even the prophet himself. Time, more than any other factor, is the grist for the mill of prophetic interpretation, and thus the justification (at least partly so) for new understandings of the Old Testament prophets in light of the Christ-event.

In the prophecies of Jeremiah and Ezekiel, Yahweh will bypass this process of speaking and listening and put his will directly into the hearts of human beings. "All the house of Israel is uncircumcised in heart" (Jer 9:26), and because of this, "Yahweh is to give his people a heart to know him" (Jer 24:7). In this verse we come face to face with the heart symbolism characteristic of the kind of salvation envisaged by Jeremiah. Indeed, one could easily regard this verse (24:7) as the new covenant prophecy compressed into one sentence. Ezekiel after him is to put the same theology in different words: "A new heart I will give you, and a new spirit I will put within you, and I will take out of your flesh the heart of stone and give you a heart of flesh" (Ez 36:26).

Jeremiah and Ezekiel emphasized the new covenant in terms of a "new heart" and a "new spirit," whereas deutero-Isaiah, another great prophet of the Babylonian exile, brings us into the presence of the suffering servant. The servant (Christ) will instill these changes in the heart, for "The righteous one, my servant, shall make many righteous and he shall bear their iniquities" (Is 53:11). One can easily imagine how Jesus may have interpreted the suffering servant oracles to his disciples and applied them to himself as they walked along the road to Emmaus:

> And he said to them, "O foolish men, and slow of heart to believe all that the prophets have spoken! Was it not necessary that the Christ should suffer these things and enter into his glory?" And beginning with Moses and all the prophets, he interpreted to them in all the scriptures the things concerning himself. (Lk 24:25-27)

As deutero-Isaiah's words gave new hope to an otherwise hopeless situation among the people in exile, so did Jesus' interpretation of the prophets concerning himself, give hope to his apostles. Paul echoes this hope which is to come into the heart (Rom 5:5; cf. 2 Cor 1:22; Gal 4:6).

Having brought attention to God's right to exercise wrath, Paul declares God's salvific justice against this background. God has the power to save, but some might ask: What is humanity being saved from? The simple answer is God's wrath. But what does this mean? Most say it is that powerful and frightening cul-de-sac of idolatrous slavery which continually blocks the way to freedom that only Christ can offer. It is a barrier brought on by the popular belief that individuals can be their own god, that somehow they can pull themselves along the road to salvation by their own efforts. Such is the endless cycle of madness known as idolatry, the human tragedy of denial and failure to recognize the real grand prize of authentic living. God's wrath boils down to leaving the unrepentant to his/her own devices. God's wrath is not lightening bolts from heaven, but what we do to ourselves (cf. Rom 2:5-8). We are our own worst enemies, it seems.

But just when everything seems hopeless there is room for hope, not because of anything humanity has accomplished on its own, but because in our endurance we have learned to look not to ourselves, but to God for salvation. Christian exultation, as always, finds its basis in hope. In this, says Paul, our expectations will not be disappointed, and we shall thus not have to bear the shame of having followed a false hope. We can be certain of this because God's love for us has already been "poured into our hearts through the Holy Spirit who was given to us" (Rom 5:5). In these terms, salvation means saving the human race from itself, individually as well as collectively. The new heart, new covenant, of Jeremiah and Ezekiel becomes for us Christ's saving act, and through the faith which can come only from the heart filled by the Holy Spirit.

CHAPTER THREE

The Head of the Body

> *They stripped him and put a scarlet robe on him, and after twisting some thorns into a crown, they put it on his head. They put a reed in his right hand and knelt before him and mocked him. saying, "Hail, King of the Jews."*
>
> (Mt 27:28-29, *NRSV*)

Not long ago I was talking with my seven-year-old granddaughter about obedience to her father. She said, "Oh, I have to obey him. He's the head honcho around my house." As I recall that brief exchange today, I wonder how the "head" of something became the present day metaphor for power and position. Is it because of the anatomical location of the head at the top-most position of the erect human body? Is it because the head holds the special senses, especially the eyes and ears, so important to our functioning as human beings? Whatever the answer, from earliest times the head of anything, human, animal, or even non-living things, has held a special dignity. Many of the special traditions and mannerism concerning the head of the body began in biblical times. Some of these we will explore in this chapter, and present to the reader for his/her consideration.

The previous chapter points out the importance of the Hebraic heart in human consciousness, reason, memory, knowledge, insight, volition, and so on. These are important functions which present-day thinking would regard as functions of the brain. Hans Walter Wolf is quick to point this out in the few instances he discusses the "head" as an anatomical unit of the body (*Anthropology of the Old Testament,* p. 46, 51). As shall be shown, however, there is a certain dignity allotted to the human head, biblically speaking, a fact which seems to indicate an instinctive understanding of its importance.

Dignity of the Head

In spite of the lack of understanding of the human brain, the head is often regarded as a part of the body deserving of great dignity. It is the site where religious leaders and kings are anointed. The anointed one, or "messiah," becomes a most important image in the Old Testament. In the New Testament the image secures its highest level of dignity and importance, since Jesus is understood as God's anointed one, the savior of all humanity. Indeed, in this new messianic understanding, the idea of kingship reaches far beyond the Old Testament image, bridging the earthly life of Jesus with the spiritual dimension of God's kingdom.

Beyond the association of the head with the act of anointing, the word "head," in itself, may be construed as possessing a measure of dignity. We speak of the head of the household just as the ancients gave to the head of anything a measure of superiority (Dt 28:13; 28:44). Even the "head of every street" (Ez 16:25; 16:31) seems a more exalted location on which to build a lofty edifice. These references, by analogy, refer to the head of the body as having greater importance than the other extremity, or "tail," (Dt 28:13; 28:44). To be the "head" of something, even for lifeless objects, is to have a position of authority, dignity and honor. The most noteworthy of this kind of reference is to the "stone which the builders rejected has become the head of the corner" (RSV, Ps 118:22). The rejected stone which becomes the "head of the corner" is also translated as the "keystone" (JB.), or "cornerstone" (NRSV). In the synoptic gospels the imagery of the rejected stone becoming the cornerstone is applied to Jesus (Mt. 21:42; Mk 12:10; Lk 20:17).

A Crown of Thorns.

A crown of thorns is placed upon the head of Jesus in mockery of Pilate's title for him as "the king of the Jews" (Mt 27:29; Jn 19:1-3). Continuing in this mode, the soldiers began to salute Jesus, crying out "Hail King of the Jews." Then they "struck his head with a reed, and spat upon him . . ." (Mk 15:18-19). Little did the perpetrators of these impulsive acts realize how powerful "the crown of thorns" would become as a future symbol of suffering and salvation.

Following his death and resurrection, Christ becomes the exalted "head of the church," while the faithful, in the context of this imagery, become the body. (Eph 1:22; 5:23). The head, therefore, develops into a powerful symbol, beginning in ancient times with the imagery of anointing and kingship (1Sm 10:1), and as stated in the prophecy of Nathan (2 Sm 7:12-17). The early church reached a new understanding of this messianic imagery, expanding the ideas of king and kingdom to the point where they exploded into a new dimension beyond space and time.

The Bowed Head.

Sometimes it is the simplest of ancient traditions that become the most meaningful to us today. Perhaps because the head of the body is often elevated in importance over other bodily structures, it is this part that is bowed in humility and obeisance. It is the tradition that explains why we bow our heads in prayer and worship, in private, or in public surroundings. In this sense, bowing the head demonstrates humility and respect in prayerfully approaching our God. Bowing of the head in prayer recognizes the unsurpassed greatness and dignity of God, compared to which we are nothing. As the apostle Paul says, we have this treasure (God's Spirit) in earthen vessels (2 Cor 4:7). Recognizing our own weakness in this prayerful symbol of humility and obeisance, we become stronger, paradoxically it seems, in the very act of bowing the head (cf. 2 Cor 12:10).

The Symbolic Head.

Certain imagery places on the head the burden of guilt, shame, even bloody reprisal. "I will requite your deeds upon your head" (Ez 16:43; cf. Nm 6:7; 1Sm 25:39), or "his blood shall be upon his own head" (Ez 33:4; cf. Jos 2:19), are examples of this Old Testament declaration. Likewise, the head can be the recipient of a malediction, or curse (2 Sm 3:29). It is appropriate here to mention the famous story of incest and revenge, told in a modern context by William Faulkner (*Absalom, Absalom*). In the biblical version, Tamar places ashes upon her head in shame, after being forced into the repugnant act by her brother, Amnon (2 Sm 13:19).

The Christian viewpoint takes this ancient cry for revenge, symbolized in the story of Tamar and her two brothers Amnon and Absalom, and gives it another slant. The gospels teach us to return the abuses of an enemy with kindness, "for by so doing," writes Paul, "you will heap burning coals upon his head" (Rom 12:20). This turnaround shows how far the early church had grown in recognizing the dangerous cycle created by violence and revenge, brutality and retaliation. This

Pauline teaching reflects the wisdom, in a practical manner, of "turning the other cheek" (cf. Mt 5:38-42).

The head, then, becomes for Christians a multifaceted symbol of suffering, humility, and obeisance. In the prayerful bowing of the head, in reflecting on our weaknesses, acknowledging human vulnerability before God, this very act becomes a fruitful source of strength and energy. In prayer we recognize, as Paul says, that "when I am weak, then I am strong," because God's power is made perfect in weakness (2 Cor 12:9-10).

Ancient Understanding of Mind

As indicated earlier, the Hebrew understanding of the human brain, the all important collection of nervous tissue located in the cranial cavity, was practically nil in biblical times. Most of the functions we presently associate with the brain were given over to the heart. The "heart," for example, had a variety of special meanings among ancient Greeks. To them it was the seat of thought as well as the seat of emotions. *Kardia* (Greek for heart), referred to the bodily organ, the one that pumps blood, but knowledge of its function, as in the case of the brain, was very limited by modern standards. Aristotle believed in the primacy of the heart as the source of "innate heat," the seat of sensation and thought, while the brain he regarded as a gland secreting cold humors to prevent overheating of the body. Hippocrates, at least, had a much better idea of the function of the brain, ascribing many functions to this organ which Aristotle associated with the heart (cf. Ralph H. Major, *A History of Medicine*).

I mention this background primarily to dispel the assumption that any functional assessment of the head (with its brain), is necessarily uniform throughout the Bible. Among the Hebrew writers of Solomon's time (10th Century BC), for example, the relationship of head and mind, or head and intellect, may have been different from what it was to become during post-exilic times.

Jews in Alexandria.

With the founding of Alexandria in 332 BC, and the influx of Diaspora Jews into this very cosmopolitan Egyptian city, one would surmise changes in the outlook of Jewish scholars regarding human anatomy and physiology. At the medical school in Alexandria, there were public dissections of the human body for the first time in medical history. This practice, which was later forbidden, did not appear again until about one thousand years later. During the time when human dissection was allowed, however, a Greek anatomist, Herophilos (circa 300 BC), made discoveries that enlightened medical sciences for centuries to come. Working in Alexandria, Herophilos described the brain with great care, noting most of the gross anatomical divisions now studied in every school of medicine. He differentiated tendons from nerves, and declared the brain to be the center of the nervous system. He is said to have divided the nerves into motor and sensory components (Major, 143), which, if true, would represent an impressive accomplishment in the absence of modern physiological techniques.

As a center of learning in Western civilization, Alexandria attracted the best of poets, philosophers, scientists and physicians. It would seem foolish to conclude that the large congregation of Jews living in Alexandia would not have been attracted to this great center of learning. During this time of intellectual growth, the Hebrew Bible was translated into the Greek language by Jewish scholars living in Alexandria (possibly charged with the undertaking by Ptolemy II Philadelphus). This version of the Old Testament later became known as the Septuagint, a translation used and referred to extensively by Diaspora Jews, and later, by early Christians.

Following the conquests of Alexander, the Greek language became the lingua franca of the Western world, and continued to be so into the first and second centuries AD, long after Roman domination had been established. It was necessary for Christian evangelists to understand and speak Greek in their travels to foreign cities. The New Testament documents, as we have them in their final form, were first written in the Greek language. It was therefore difficult to escape the influence of Hellenism (Greek culture and language) in those days, especially among merchants and educated peoples.

Mind and Brain.

The modern reader of the Bible may be mislead by many translations of the Hebrew text in which one encounters the word "mind." In a large number of cases, the Hebrew word is either leb or lebab, which a literal translation would render as "heart." In the few instances in which "heart" and "mind" are used together in the same verse (1 Kgs 8:48; 1 Chr 28:9), the "mind" here may refer to something akin to, "life," or perhaps, "soul" (Heb., nephesh). The point to be made, regardless of the literal meaning of the word, is that the ancient Hebrew writer was not likely to think in terms of the human brain as having anything to do with the mind, wisdom, or the intellect.

There may be exceptions to this general proposition regarding the mind-brain relationship, but it would be difficult if not impossible to prove. The closest approximation of some level of understanding of the mind-brain relationship can be found in the book of Daniel, written sometime around 165 BC, well after Greek learning had established a firm hold on Jewish thought. In fact, the author of Daniel wrote during the persecution of the Greek leader Antiochus Epiphanes. The author speaks of dreams and visions as being associated with the head (Dn 2:28), a relationship not found in earlier, pre-exilic writings. This notion of the head with dreams and visions would seem to reflect a more sophisticated view of brain function in the book of Daniel. If so, such a development would have been a consequence of the greatly enriched scientific understanding during this period of Greek influence.

The Greek words for mind, thinking, or inner self, etc. (e.g., *nous, phronema, psyche*) increased over time, especially among New Testament evangelists. That is not to say that *kardia* was replaced with one of these words for "mind." There was, however, a definite tendency to use "mind," or one of its cognates, in a sense that overlapped an accepted meaning for the Hebraic heart (see my book, The Biblical Heart, for details). According to Fitzmyer (NJBC), kardia (heart) and nous (mind) become practical equivalents in Pauline theology.

The Face of the Head

The word, "face," whether expressed in Hebrew, Greek, or English, covers a broad range of meanings. The Old Testament speaks of the "face of the earth," or the "face of the waters," the "face of man," even the "face of God," using the same Hebrew word. In spite of multiple applications for this one word, there is a common thread that connects all of them. To see the "face" of anything, regardless of the word's inanimate, human, or divine reference, indicates the recognition of identifying features of the object. One recognizes a body of water by observing its face, used in this case as a synonym for "surface." One does not need to chemically analyze a pool of water to know what it is. In most cases, by simply observing the (sur)-face of anything such as the ground, or water, or the face of a human head, we recognize what it is. That is because the surface (or face) of most things disclose what lies beneath it as well, within certain limits, of course.

The Face that Intensifies.

In the Scriptures, as we move from identifying the face of inanimate objects to the human face, there is usually a consequence of the encounter that goes beyond mere recognition. A simple word study of the biblical text makes it abundantly clear that pronouns, such as "your," "his," "her," are usually supplemented by the word face. "When you see his face," for example, or "you shall not see my face again," are common expressions. In the latter example, instead of saying simply, "you will not see me again," the absence is intensified by not seeing "the face" of the person. A human face may tell us not only who the person is, because we may also identify the person as a relative, or a close friend, with all the incidental and cumulative knowledge about that person at our disposal (cf. Gn 48:11). That is why the statement, "you will not see my face again," seems to carry with it a sense of reality, or personal impact, not implied in the use of the pronoun, "me," used alone. The human face, therefore, includes a depth of meaning not implied in the face of an inanimate object. In the case of a human, especially among relatives or loved ones, "face" means much more than "surface."

There are many other examples of how the face intensifies a given human situation or condition. To bow one's head, as mentioned earlier, is a sign of respect, or in some cases, of humility and obeisance. To fall on one's face is to demonstrate the same emotion in the extreme (Gn 17:3; Nm 22:31; Jos 5:14; 7:6; Ez 1:28). In a sense, the person who falls in this manner identifies with the earth beneath his face in abject humility. Contrast that position with the person being shown the obeisance (a king, perhaps), who stands with the earth beneath his feet.

Other examples of this intensification include, to meet someone "face to face," or "to hide one's face," or to "set one's face" against another. Other expressions are the "hardened face," the "turned away face," the "covered face," and the "confused face." We find these particular uses throughout the Bible, in Old and New Testaments, almost all of them supporting the idea of bringing a greater depth of emotion to the reality of a given situation.

To Set One's Face.

To "set" one's face indicates a determination that is unwavering. This can usually be inferred from the context of the biblical passage. It is a stronger indication of what is about to happen than a simple decision would indicate. Thus, to "set" one's face is to make a decision which is not likely to be reversed. For example, the gospel of Luke could have reported Jesus as having decided to go up to Jerusalem. Instead the passage states that "he set his face to go to Jerusalem" (Lk 9:51), indicating a level of determination that was steadfast.

The Face in Translation.

Translations are of extreme importance when it comes to understanding these nuances of bodily reference. In the gospel of Mark (1:2), in reference to the role of John the Baptist as one who will "prepare the way," two popular translations, read, "I am sending my messenger ahead of you" (*NRSV* and *NAB*), whereas the Greek says, "I am sending my messenger before thy face, who will prepare thy way." The Greek word for face (prosopon) is simply deleted from the above English versions. Omitting "face" from the translation appears to soften, if not totally eliminate, the intent of the evangelist to announce the determination of God to intervene in history in the strongest possible way. My remarks, here, should in no way be construed as criticism of these two English versions. Translators are constrained by a difficult balancing act in their work. They must decide where and when to compromise the literal sense for a more understandable, modern text. By the same token, and in the absence of Hebrew and Greek proficiency, a more literal translation, such as The Revised Standard Version, is more useful in word studies such as presented here.

The Face as Whole Person.

In many cases, the face may heighten the drama of a circumstance, especially when it is seen to represent the whole person, deity, etc., a common literary device known as synecdoche (metonymy). "A man of understanding sets his face toward wisdom, but the eyes of a fool are on the ends of the earth" (Prv 17:24; cf. Jer 51:51). In this example it is not just the man's face that seeks wisdom. The face in this case, rather, represents the whole person.

Speaking of God in terms of some anatomical structure, such as "the face" or "the hand," is also a common extension of synecdoche. In the poetry of the Psalms, seeking the face of God (Ps 11:7; 27:8), or beholding it (Ps 17:15; 42:2), or to have it shine upon you (Ps 31:16), certainly deepens and intensifies the sense of what the psalmist is trying to convey. Attributing anatomical structure to God, commonly in the psalms (the practice of anthropomorphism), has its advantages as a literary device. To provide a "face" or a "hand" to an otherwise abstract, transcendent being, must have given the psalmist an emotional hold and a feeling of close relationship to the God of Israel.

Paul's Use of Face.

Some of the best examples of the face as a device for intensification are found in the writings of the apostle Paul. Comparing our earthly existence with what is to come, Paul says, "For now we see in a mirror dimly, but then face to face. Now I know in part; then I shall understand fully..." (1 Cor 13:12, my italics). Thus, to see the reality of what is to come, to fully understand the afterlife, we must be there and see it "face to face." In our present, earthly existence, we can see reality only dimly. Indeed, says Paul, "...no eye has seen, nor ear heard, nor heart of man conceived, what God has prepared for those who love him" (1 Cor 2:9).

I am struck by the similarity of this analogy (the dim mirror) to the shadow world of Plato's *Republic.* Plato explains humankind's existence among the living as akin to a group of people seeing shadows on the wall of a cave. These shadows are all they can see, since they are confined to the cave of human existence. The true reality is outside the cave, according to Plato, consisting of that which is the source of the shadows, rather than the shadows themselves. Paul, born in the cultured city of Tarsus in Asia Minor, was probably aware of Plato's work. As many scholars believe, Paul would certainly have had the opportunity to become steeped in Greek philosophy and rhetoric. If such were the case, it would seem natural for him to apply some of these concepts to explain the Christian view of life after death.

With Unveiled Face.

Another, perhaps more famous example of Paul's use of the word "face" relates to his comparison of the face of Moses and that of the followers of Christ. The veil which covers the shining face of Moses (cf. Ex 34: 30-35), is the basis of the comparison. Philo (*On the Life of Moses*, II, 70), understands the event as follows:

> [Moses] descended again forty days afterwards, being much more beautiful in his face than when he went up, so that those who saw him wondered and were amazed, and could no longer endure to look upon him with their eyes, inasmuch as his countenance shone like the light of the sun.

Paul has another interpretation of the veil on the face of Moses. He sees the facial covering as lying between the glory of God and the Israelites, and explains how they could fail to see the "end of the fading splendor" (2 Cor 3:13). But now, continues Paul, "when a man turns to the Lord, the veil is removed" (2 Cor 3:16). Paul thus arrives at one of his most powerful and dramatic conclusions, that now, "we all, with unveiled face, beholding the glory of the Lord, are being changed into his likeness from one degree of glory to another; for this comes from the Lord who is Spirit" (2 Cor 3:18). This remarkable passage refers to a transformation, a metamorphosis, in which Paul sees glory heaped upon glory until our lowly body will change "to be like his [Christ's] glorious body" (cf. Phil 3:21). These images (e.g., 2 Cor 3:18; Phil 3:21), says Fitzmyer (*NJBC*, 1401), led the Greek patristic writers to derive the idea of theosis or theopoiesis, the gradual "divinization" of the Christian.

To the extent that such a "divinization" is possible, the accretion of glory that comes to the Christian is not simply because the person has faith, but rather because he/she participates in the Christ-event. The actualization of the Christian transformation, therefore, occurs as the believer "partakes" of Christ's death and resurrection (cf. Keck, 75, et seq.). "For if we have been united with him in a death like his, we shall certainly be united with him in a resurrection like his" (Rom 6:5).

In light of these theological insights, the "unveiled face" becomes a powerful symbol of the transformation brought about through faith and its consequences. We are no longer simply "earthen vessels" because we possess a treasure from the Spirit (2 Cor 4:7). We are, in light of this treasure, "a new creation; the old has passed away, behold, the new has come" (2 Cor 5:17; cf. Gal 6:15).

Paul did not need to understand modern biology, entropy, or any of modern physics to know and understand the basic "laws" of nature. When he preached about a "new creation," Paul was talking absolute heresy as far as the laws of nature are concerned, *and he probably knew it.* The body (*soma*), as common knowledge in Paul's time, and also in the view of Greek science of those days, may have added up to more than the sum of its fleshy parts, but only in a functional sense. Beyond this miracle of nature there was no principle (with the possible exception of magical beliefs) to account for the kind of transformation of which Paul speaks, excepting a divine one. The power of God, shining gloriously upon us, through the spirit of Christ (2 Cor 3:18), is the force Paul sees as constituting this divine principle which God shares with believers. "Being changed," in the Apostle's thinking, is to be "divinized," to share in God's self-communication, which is to share in God's divine nature.

The Scriptural Importance of Face.

In regard to the "face," Wolff asserts (p. 74) that a "man's face is far more important that his 'Head,'" at least in the Old Testament. One reason Wolff make this assessment is because the face reminds us of the manifold different ways in which "a man can give his attention to his counterpart." Events are reflected in the features of the face, and, of course, the organs of communication, eyes, mouth, and ears, are gathered together here. Among all the organs and limbs, Wolff asks, "is it not here that we ought to come close to finding out what man's being consists of, and what distinguishes him from all the other beings?"

The face is important for all the reasons Wolff has advanced. It should not, however, be construed as "more important" than the head, since the face is a part of the head, just as are all its other features, including the brain, sense organs, and crown of the head. The eyes, ears, and oral structures (mouth, palate, throat, etc.), will be considered in subsequent chapters, since each of these organs is important enough in its own right to deserve individual treatment. It is also true, of course, that there is critical interdependence of body parts, as the apostle Paul so eloquently points out:

> Indeed, the body does not consist of one member but of many. If the foot would say, "Because I am not a hand, I do not belong to the body," that would not make it any less a part of the body. And if the ear would say, "Because I am not an eye, I do not belong to the body," that would not make it any less a part of the body. If the whole body were an eye, where would the hearing be? If the whole body were hearing, where would the sense of smell be? But as it is, God arranged the members in the body, each one of them, as he chose. If all were a single member, where would the body be? As it is, there are many members, yet one body. (*NRSV*, 1 Cor 12:14-20).

Paul is here speaking, by analogy of course, of the various gifts that are unified in the one Spirit, such that the faithful participate as individual members of the body of Christ (1 Cor 12:27). In subsequent chapters I will examine separate parts of the body, hopefully without neglecting their interrelationships. My goal will be to highlight the importance of these structures in the minds of the Hebrew and New Testament authors in communicating effectively with their respective audiences.

CHAPTER FOUR

The Eye

> *O Lord, all my longing is known to you; my sighing is not hidden from you. My heart throbs, my strength fails me; as for the light of my eyes—it also has gone from me*
> (Ps 38:9-10).

"As the eye sees, so the mind thinks," or is it the opposite, "as the mind thinks so the eye sees?" It doesn't matter, really, because both are true to some extent. The important thing about the special senses, especially the eyes and ears, is that they are the entry portal for all we know. Everything we learn comes to us through the senses, from the time we are born until we die. Even abstract notions, such as God, eternity, spirituality, are arrived at from examining the world through the senses. In the absence of the senses a person would be little more than a vegetable insofar as the ability to learn and live like other human beings. Actually, one would not even make a good vegetable in the absence of sensory experience of some kind. Vegetables in the wild grow to maturity and reproduce their own kind, something a sensory deprived person could never do.

The point of this discussion is to bring out how important the senses are, along with the nervous system which receives the sensory impressions (brain, spinal cord, and peripheral nerves). Oh, sure, other mammals and vertebrates have a nervous system as well, but the highly developed one possessed by human beings is what distinguishes us from the lower forms. It is our highly developed brain that allows us to analyze sensory input coming from the eyes, ears and skin, and make it into useful experiences beyond what any other animal can do with the same information. We not only learn from sensory experience, but we reflect on it, making it a part of our unique personality. We ask questions of this sensory input that go beyond what we see and hear. Why is the sky blue? What are clouds and stars? Why does water boil? What makes water solidify into ice? Why are we here? How did this world we live in come about? Why is there something rather than nothing?

These kinds of questions define us as distinctly human, because they take everyday experiences and ask for explanations that transcend the ordinary. But here is the catch. Some individuals can be so intent on the pleasurable aspects of life, that they never reach their full potential as human beings. These individuals prefer not to think about abstract ideas, moral values, or virtuous living, because such thoughts would curtail their ability to enjoy their carnal impulses. Biblically speaking, to the extent that we indulge in self-centered living at the expense of higher ideals, we diminish our awareness of God, and in so doing we become *less human*. At the very least, those who insist on living in a dissolute manner diminish their human authenticity. That is because these individuals insist on using the senses as a lower animal would do, instinctively, with little regard for anything beyond the satisfaction of biological urges.

These are the persons of whom the prophets say, "they have eyes that do not see, ears that do not hear." The comforting thought, however, since we are all sinners, is that God loves us in spite of our shortcomings. God keeps calling to us, trying to heal the blindness of our eyes and to open our deaf ears.

This chapter will be divided into two parts, "**The Seeing Eye**" and "**The Blind Eye**." The eye is one of the most important of all metaphors in the Bible for the division of humankind into those who see, and those who are sightless. Strangely, Hans Walter Wolff (Anthropology of the Old Testament) hardly mentions the "eye" and "seeing," while ascribing much greater importance to the "ear" and "hearing." I shall examine both of these sense organs in some detail and let the reader decide if one sense, biblically speaking, is more important than the other. The chapter following this one will be devoted to the ear, giving equal attention to the "ear that hears" and the "deaf ear."

The blind person of the Bible is not necessarily one who is physically blind. There are biblical characters who are physically blind, of course, but an important distinction must be made between the "spiritually blind," and the actually sightless, or blind person. Sometimes it is obvious from the scriptural context, that a certain individual is physically, not spiritually, blind. In other cases the distinction is not so clear, and one must examine the text closely to determine which is which. Other times, particularly in the New Testament, spiritual and physical blindness are contrasted with one another. In any case, it is helpful to know the symbolic history of the eye as recorded in the Old Testament before making a decision concerning a particular passage of "seeing" or "not seeing" in the New Testament.

The Seeing Eye

This section deals with the eye as a physically sound organ. That is not to deny, of course, that some degree of spiritual blindness is always present. Spiritual blindness is part and parcel of being a mortal human being.

Lifting Up the Eyes.

One of the most common of biblical phrases concerning the seeing eye is, "he lifted up his eyes." Modern translations such as *NRSV*, or, *NAB*, may translate the Hebrew simply as "he looked up" (cf. Gn 13:10; 18:2; 24:63). In either case, the person does something ("raises the eyes," or "looks up"), prior to the act of focusing the eyes on some object. The expression, I believe, is significant, especially since it is found throughout the Old and New Testaments. Abraham "lifted up his eyes" and beheld three men standing before him (Gn 18:2). The three turn out to be Yahweh and two of his messengers. Joshua "lifted up his eyes" and beheld a man with drawn sword. The "man" identifies himself as the "commander of the army of the Lord" (Jos 5:13-14). The psalmist writes, "I lift up my eyes to the hills," asking, "from where will my help come?" In answering the question, the psalm reads, "my help comes from the Lord who made heaven and earth" (Ps 121:1-2). An oracle of second Isaiah exhorts the people to "Lift up your eyes to the heavens, and look at the earth beneath; for the heavens will vanish like smoke, the earth will wear out like a garment, and those who live on will die like gnats; but my salvation will be forever, and my deliverance will never be ended" (Is 51:6). All in all it seems an apt conclusion that the "lifting of the eyes" is saying, "*face reality*," or "*recognize that God is present in your life and act accordingly*."

In the gospel of Luke, Jesus "lifts up his eyes" and begins to speak the beatitudes, the sermon on the plain (Lk 2:30). Following the transformation of Jesus in the gospel of Matthew, when the disciples "lifted up their eyes," they see Jesus in a totally new light (Mt 17:8). They begin to understand, as never before, Jesus' title for himself as

"Son of Man."

Lifting up the eyes can exhort the people not just to face the reality of the Lord's presence in their lives, but to wake up to the evil they do. As the prophet Jeremiah, writes, "Lift up your eyes to the bare heights, and see! Where have you not been lain with? By the waysides you have sat awaiting lovers like an Arab in the wilderness. You have polluted the land with your vile harlotry" (Jer 3:2). In the parable of the "Rich Man and Lazarus" (Lk 16:19-31), the wake up call (to spiritual awareness) is ignored at first. When it comes again it is too late, because the rich man who had mistreated Lazarus dies and resides in the torment of Hades. He "lifted up his eyes and saw Abraham far off, and Lazarus in his bosom" (v. 23). In lifting up his eyes the rich man saw the reality of his past sins, recognizing also what he had lost in his lack of compassion.

In the gospel of John, Jesus asks the disciples, "Do you not say, 'there are yet four months, then comes the harvest.' I tell you, lift up your eyes, and see how the fields are already white for the harvest" (Jn 4:35). In this case Jesus is warning the disciples that there is a kind of harvest that goes beyond the harvesting of grain and food. There is an eschatological harvest that is more important, because in this one, eternal life is at stake. In other words, Jesus is exhorting them to see beyond the here and now, to be aware of a truth that escapes the present dimension of time and bodily needs.

There are many other examples of "lifting the eyes." I have mentioned just a few selected ones in an effort to make a point. To me, "lifting up the eyes" has a significance that is not implied by the action of simply, "looking up." Indeed, "lifting the eyes," in almost every instance of its use in the Bible, presages some important event, action, or in some way prepares the reader for the recognition of a momentous reality. Usually the event has to do with a truth that relates, directly or indirectly, to divine reality, such as, *the Lord, our God, is present to us in history and we should act accordingly.*

Other Common Expressions.

Many expressions used in connection with the eye are too numerous to cover in detail here. In most instances the meaning is fairly obvious from the context and needs no further clarification. Examples of such adjectives are "greedy," "weeping," "haughty," or "envious eyes," eyes that "grown dim" with age, eyes that "flash" or "blink," and so on. "Painted eyes" (Jer 4:30; Ez 23:40) may refer to vain affectations, or be symbolic of such behavior. The "winking eye" may indicate a puffed up sense of self-knowledge (Ps 35:19) or tacit approval of some misdeed (Prv 10:10; 16:30). Proverbs makes a number of references to the eyes in the context of wise sayings or aphorisms which give advice, or warn against certain behavior. These are important, but are, for the most part, self-explanatory and need no further elaboration here.

Other expressions are familiar to us in our own day, such as, "In his or her eyes," which may be construed to mean, "in his or her opinion." "To have eyes" for something or someone is to desire or covet that something or someone. An extension of this, to be "wise in your own eyes" (Prv 3:7; 26:5; 28:11), or "pure in your own eyes" (Prv 30:12), is a warning against self praise. Job's friends believe him to be "righteous in his own eyes" (Jb 32:1), and question him no more. Here the implications of Job's self-righteousness are more complex than at first glance. This wonderful folktale introduces profound questions about suffering, good and evil, and more precisely, "why do bad things happen to good people?" What kind of God is our Creator and Lord to allow such things to happen. In his suffering, immersed in a troubling and difficult turn of events, Job has the courage to ask God:

> Do you have eyes of flesh? Do you see as humans see? Are your days like the days of mortals, or your years like human years, that you seek out my iniquity and search for my sin, although you know that I am not guilty, and there is no one to deliver out of your hand? (*NRSV*, Jb 10:4-7)

The important and most powerful question Job asks is, "Do you have eyes of flesh," because this condenses all the questions and statements to follow. In other words, what kind of God are you, Lord? Do you even care what happens to mortal flesh? Are you the just and impersonal, yet powerful creator of all? In asking all these things, Job assumes a kind of reciprocal relationship between God and mortal flesh. In speaking to God in such a manner he displays great hope and faith that he will receive, in some manner or other, answers to his questions. How wonderfully this story encapsulates the human condition, the anguish, confusion, distress, hopes and prayers of all peoples in every civilization down through the ages.

The Blind Eye

Here I will consider one of the most important metaphors in the entire Bible. In the juxtapositioning of physical blindness with various kinds of sinfulness, one finds a recurring literary device commonly found in both Old and New Testaments. Spiritual blindness is comparable to another powerful metaphor, and that is, "hard heartedness." In the case of blindness, the biblical writer's intentions are not always clear in this regard. Often physical blindness is intended, but there are other times when spiritual blindness alone is to be construed. Additional examples imply that both physical and spiritual blindness exist together, such that the healing of the physical malady leads to, or is a consequence of, a spiritual awakening. The purpose of the following discussion is to help solve this problem insofar as that is possible.

There is no question that the prophet Isaiah is speaking of spiritual blindness in the oracle dealing with perversity (Is 29:9-16). Following those words, the prophet moves into a theme of redemption, in which "the deaf shall hear the words of the scroll, and out of their gloom and darkness the eyes of the blind shall see" (Is 29:18; cf. 30:20; 32:3). Isaiah is not always clear, however, on the issue of spiritual vs. physical blindness, which may be, strange as it may seem to us, an intentional literary device. Whereas ambiguity is certainly no virtue to the modern scientist, this is not necessarily the case among ancient Hebrew writers. In other instances the nature of the "blindness" is made clear, for as Isaiah's oracle of promise and redemption says, "Bring forth the people who are blind, yet have eyes, and the deaf, yet have ears (Is 43:8; cf. Jer 5:21).

In certain instances, Isaiah and Ezekiel seem to draw parallels between idols and those who worship them. That is, for those who look upon blind idols who can neither see nor hear, they too will become blind in their idolatry. Thus, the idol worshiper becomes like the piece of wood he worships, blind and deaf (cf. Ez 6:9; Is 44:18).

The Opened Eye.

The "opened eye" is an example of a phrase that is significant, but there is a diversity in translations of the Hebrew word which means, in a literal sense, "to open," in this case, to "open the eyes" (cf. Stong's Hebrew dictionary [#8365]). The problem is illustrated in Numbers 24:3, in which the Hebrew word for "open" is translated in different ways (RSV, "opened eye;" NAB, "true eye;" NRSV, "clear eye"). For consistency with the Hebrew meaning, the more literal translation of the "opened eye" will be used here.

In the Numbers story, Balaam, a soothsayer, is asked by Balak, son of Zippor, king of Moab, to put a curse on Israel. But it was revealed to Balaam that the Lord was pleased with the Israelites and had blessed them. The spirit of God came upon him, and he uttered this oracle:

> The oracle of Balaam son of
> Beor,
> the oracle of the man whose
> eye is opened,
> the oracle of him who hears
> the words of God,
> and knows the knowledge of
> the Most High,
> who sees the vision of the
> Almighty,
> falling down, but having his
> eyes uncovered:
>
> (Nm 24:3-4)

Here it is obvious that the phrase, "the eye is opened," makes no reference to physical blindness. Balaam's eyes were opened in a spiritual sense. Because of his opened eyes he was able to see the will of God, whereas before he was blind (and deaf) to God's word. Other examples of the "opened eye" may not be so clear.

Isaiah (35:5) relates that the "eyes of the blind [in paradise] shall be opened." In the first Servant Song of second Isaiah (Is 42:7), "the eyes that are blind" are opened, and the prisoners will be brought from the dungeon. In the first instance, will not all the chosen people in paradise have their eyes opened to the glory of God? That is not to say, of course, that the sightless person who finds himself in paradise will not also be cured of the physical malady as well. In both oracles the nature of the "blindness" is uncertain, but the context suggests that the blindness is spiritual, not physical. In the Servant Song, many scholars believe the "servant" represents the nation of Israel, while others regard him as an individual, or perhaps a combination of these. The major emphasis, here, as in the first instance (Is 35:5), is in reference to spiritual blindness, because it does not make sense that an entire nation would be physically blind.

Among the wisdom books, there are at least two instances of the eyes being opened (Jb 14; Ps 146:8). In Job the reference is to God's eyes (an anthropomorphism), and thus has nothing to do with physical blindness. In the reference to the Psalm 146, verses 7-8 read, "The Lord sets the prisoners free; the lord opens the eyes of the blind. The lord lifts up those who are bowed down; the Lord loves the righteous." The probable post-exilic setting of this psalm might be a clue as to the meaning of the psalmist when he writes about setting prisoners free and opening the eyes of the blind. One might ask the legitimate question: "Were there so many blind people in those days as to mention such a miracle in the same breath with freeing prisoners?" In the same verse, the Lord lifts up those "who are bowed down," an obvious reference to the oppression of Israel in Babylon (perhaps including the oppressed in other places as well). The psalmist's context here relates to a number of prisoners and oppressed people freed through the power of God. The blindness referred to is certainly spiritual rather than physical.

The prophets, for the most part, explain the destruction of Jerusalem and the exile of the Israel as Yahweh's wrath over the sinfulness (idolatry and hard heartedness) of the people (see the oracles of Jeremiah and Ezekiel). The major problem with the nation of Israel, therefore, was spiritual blindness, and not some kind of physical sightlessness. One cannot rule out the possibility of the psalmist (Ps 146:8) having reference to physical blindness, but here again, the context mediates against such an interpretation. There are other examples in the Old Testament which refer to "eyes being opened" (1 Kgs 8:52; 2 Kgs 6:17; 6:20). In all these cases the context leaves no doubt that physical blindness is not a factor in the interpretation of these verses.

In the gospel of Matthew, as Jesus is leaving the house of the official whose daughter he raised, two blind men followed him, "crying loudly, 'Have mercy on us, Son of David'" (Mt. 9:27-30). In this case and also in one other healing of two blind men (Mt. 20:33-34), the Greek word, *anoigo*, meaning "to open," is the request of the blind men in regard to their eyes. Also, in both instances, Jesus "touches" their eyes and their sight returns. Matthew makes no effort to explain how, in the first instance, the blind men were able to "follow him." The author does not say they were led, or by what manner the two were able to follow him in their sightless condition. Perhaps they followed the sounds of the crowd, or grabbed hold of a sighted person's garment, but this is not explained in the gospel account. In the second case (Mt. 20:33-34), the blind men were "sitting by the roadside." I mention these circumstances merely to contrast them with other gospel accounts of blind healings in which the evangelist **does** make it clear that the blindness is physical. Otherwise, there are several obvious examples of spiritual blindness found in the gospel of Matthew, most of them clustered together in chapter twenty-three (Mt 15:14; 23:16; 23:17; 23:19; 23:24; 23:26).

There is a blind man story in the gospel of John which occupies the whole of chapter 9. In this case the man is said to be "blind from birth," obviously in reference to a physical anomaly. The narrative is an important one, since it uses physical blindness in contrast with the spiritual blindness of the Pharisees, who question the once afflicted man. The man blind from birth becomes enlightened spiritually as well as physically, while the Pharisees are unable to properly interpret the significance of Jesus' sign (the healing miracle), and remain spiritually sightless (unable to see God's will).

In the gospel of Mark, Jesus takes great pains, using the blindness metaphor, to refer to his own disciples. After they do not understand, he asks, "Having eyes do you not see, and ears do you not hear? And do you not remember?" (Mk 8:18). When the disciples continue in their lack of understanding, the gospel writer makes sure the reader understands the contrast between physical and spiritual blindness. In the next pericope following his rebuke of the disciples, Jesus and the disciples came to Bethsaida where some people "brought to him a blind man, and begged him to touch him" (Mk 8:23). Jesus "took the blind man by the hand, and led him out of the village." The first time Jesus tried to heal the man (when he had spit on his eyes and laid his hands upon him), it didn't work. The man said "I see men; but they look like trees; walking" (8:24). So Jesus laid hands on the man once again, whereupon, he "saw everything clearly" (8:25).

One can only conclude that the author of Mark had a reason for bringing in all the extraneous material about men who appeared as "trees walking." The gospel also makes it clear that "some people brought to him" a blind person. But Jesus did not heal the man on the spot, "he led him out of the village." If that were not enough to convince the reader that the man was truly blind, Jesus is reported to need two attempts at the restoring the man's eyes. There seems little question here that the narrative is making it entirely clear to the reader that this man's blindness is a physical malady.

The arrangement of the blind man narrative here is most important. Following this dramatic healing, Jesus went on with his disciples to Caesarea Philippi. On the way he ask them, "who do men say that I am?" (Mk 8:27). They told him "John the Baptist; others say Elijah; and others one of the prophets." Then he asked them, "But who do you say that I am?" In answer Peter provides the only response, saying "you are the Christ." The initial conclusion of the reader, namely, that Peter indeed understands what Jesus is all about, is quickly dispelled in the following verses (31-33). Hearing Jesus tell them of his coming passion, death and resurrection, Peter rebukes him, whereupon Jesus says to Peter, "get behind me Satan! For you are not on the side of God, but of men."

Two other passion predictions follow (Mk 9:30-50 and 10:32-52), and each time the disciples continue in their blindness as to the will of God concerning Jesus. Then comes the second healing of a blind person, the blind Bartimaeus (10:46ff). This time the blind man is a beggar, a livelihood one would expect of someone truly blind. Again, it seems fairly obvious that the author of Mark is trying to give the reader a message. By the construction of the narrative, placing a blind man at the beginning and at the end of the three passion predictions, it seems the reader is being asked to compare and contrast physical blindness with the obduracy (or spiritual blindness) of the disciples.

Intended or not, the comparison goes even deeper. To be sure, the blind men episodes imply the ease with which God may cure a physical difficulty such as blindness. It is less likely, however, for God to force someone into accepting him as Lord, and certainly not without compromising human freedom in this regard. Christians believe that the act of loving God involves not only the freedom to love, but the freedom not to love, or even to do evil. God cannot force anyone to love him, for to do so would be tantamount to recreating humanity as automatons. This is not to say that the disciples did not love Jesus, because all indications are that they did. But they did not fully understand, and would not come to an understanding until after the death and resurrection of Jesus. God's strategy was for this final understanding of the Messiah to come in its own time. Even then, a free will act is required in order to come to grips with the this difficult reality. As Paul said, the crucified Christ is a stumbling block to the Jews and a folly for the gentiles (1 Cor 1:23; cf. 1 Cor 1:18; 1:21).

Distinguishing physical from spiritual blindness can be crucial to understanding words the evangelists place on the lips of Jesus. The following is an example of a pericope in which the text speaks of a "sound eye," but does not refer to physical blindness.

> Your eye is the lamp of your body; when it is not sound, your body is full of darkness. Therefore be careful lest the light in you be darkness. If then your whole body is full of light, having no part dark, it will be wholly bright, as when a lamp with its rays gives you light. (Lk 11: 34-36, *NAB*)

Christianity would be a cruel and unfeeling religion if it were to interpret the above as a pronouncement against a physical malady. This is not the case, of course, for we know from numerous other textual examples how Jesus is depicted as having profound compassion for the sick and disabled.

The meaning of the darkness/light metaphor is made even clearer when the apostle Paul describes to king Agrippa his conversion and his calling to preach to the Gentiles. That is, "to open their eyes, that they may turn from darkness to light and from the power of Satan to God..." (Acts 26:18). Obviously, the whole Gentile population is not afflicted with physical blindness. One might contrast this with the description of Paul's blindness following his conversion experience. Here, physical blindness is certainly implied, for "they led him by the hand and brought him into Damascus" (Acts 9:8). And yet, one could argue that the author of Luke-Acts is making a point as to Paul's spiritual blindness as well, since it relates to an awakening to the reality of Jesus as Christ and Lord. It could be said that Paul was so confused in the overwhelming power (and brightness) of the theophanous event, that he was temporarily blinded. In other words, what Paul experienced was the *blinding truth.* I am not saying that this was the intended meaning of the narrative, but such an interpretation does seem to add another dimension to Paul's conversion experience.

Another thought to bear in mind comes with Paul's reminding us that there are limits to what "the eyes" are capable of seeing, or are allowed to see. Thus, some level of spiritual blindness is part and parcel of being a mortal human. In first Corinthians (2:9) Paul's audience is told, "What no eye has seen, nor ear heard, nor the heart of man conceived, what God has prepared for those who love him." This paraphrasing of Isaiah (64:3) certainly sets limits on the faithful's ability to see their future destiny, that is, what is in store for them after death. Yet, it was these "visual" limits that provided encouragement to the early Christian martyrs.

The letter from the church at Smyrna to the church at Philmelium, known as the "Martyrdom of Polycarp," is the oldest known document that describes a Christian martyrdom outside the New Testament. Because of its belief in one God, the early church found itself at odds with the Roman state which permitted no compromise (Lord Christ versus Lord Caesar) and from which only one of the two parties would emerge victorious. In the eyes of second century Rome, therefore, Christians were disloyal atheists who threatened the well-being of the empire (Holmes, 222).

There is some dispute over the exact date of Polycarp's death, but it probably occurred close to AD 156. The text of the document in which the martyrdom is described may have been written by eyewitnesses not long after the event (15:1). One excerpt is as follows:

> And the fire of their inhuman tortures [the victims were burned to death] felt cold to them, for they set before their eyes the escape from that eternal fire which is never extinguished, while with the eyes of their heart they gazed upon the good things which are reserved for those who endure patiently, things "which neither ear has heard nor eye seen, nor has it entered into the heart of man," but which were shown to them by the Lord, for they were no longer men but already angels (Mar 2:3; cf. 1 Cor 2:9).

CHAPTER FIVE

The Ear

> *And Aaron's sons were brought, and Moses put some of the blood on the tips of their right ears and on the thumbs of their right hands and on the great toes of their right feet*
> (Lv 8:24)

There is an adage about those who "march to their own drummer," or "dance to their own tune." In a twisted kind of way this might summarize what this chapter is all about. Suppose you entered a dance hall and saw a large group of people dancing. There was another group milling about, apparently with no interest in participating in the dance. That was because the inactive group could not hear the music. In fact, there was a beautiful melody seeming to emanate from the rafters, but there was no musician visible, no banjo being plucked, no sweaty fiddler playing his heart out. Since there was nothing coming to the ears which the indolent group considered to be music, they did not, or could not, participate in the joyful celebration of the dance.

Another small knot of people was very outspoken in their criticism of the dancers. They laughed, pointed fingers, and remarked at their stupidity. Actually, these outspoken critics had made an attempt at dancing. They went out on the floor and pretended to hear the music and even attempted to dance. They looked around, hoping to receive approval from the other dancers. Eventually they bumped into each other, slipped down and fell upon the hard surface of the dance floor. The others continued to dance, but reached down and tried to help those who had fallen to regain their feet. Instead of receiving thanks for their efforts the dancers got only angry responses from those who had fallen. "Take your hands off me," they yelled.

Eventually the pretenders became infuriated with those who kept on dancing with such skill, because they could not do the same. They left the dance floor in disgrace, then began to laugh and point their fingers at the group still dancing. "How stupid they are," they shouted. "Look at them out there, pretending to hear the music when there is none to be heard." The more beautifully the dancers moved, the more outlandish were the insulting cries of the ones standing on the sidelines.

Does all this sound familiar? Doesn't the above scenario make a wonderful metaphor for those who laugh and make fun of religious devotion? The critics think they know all about religion because they tried it and it didn't do anything for them. The dancers, of course, are those who hear the word of God and do it. As the prophets say, let those who have ears to hear, listen to the word of God. There has always been physiological deafness, of course, but in this case I am talking about spiritual deafness.

As promised in the previous chapter, I will discuss the ear in the same manner as the eye, first, under the heading, "Ears that Hear," followed by the "Ears that Do Not Hear." As in the case of "the eye," it is not always clear if the biblical text refers to a physical difficulty or a kind of spiritual obduracy. Sometimes the two conditions are contrasted, perhaps to emphasize a theological point. The term "spiritual blindness" shifts in this chapter to the idea of "spiritual deafness." These terms are similar to, if not identical with the condition of "hard heartedness," all of which indicate a lack of receptiveness to the will of God. Such spiritual obduracy is a recurring theme and complaint from the mouths of prophets and psalmists alike. Indeed, all the books of the Bible make reference to this hardness, which at times seems to be a constant companion of the people. Spiritual blindness and deafness have always been humankind's greatest failing, right down to the present time. In this chapter God brings his message of love, hope, friendship and comfort to those who "have ears to hear." The problem comes, biblically speaking, when the people "have ears, but do not hear."

Ears that Hear

As indicated earlier, Old Testament verse often uses personal pronouns in association with anatomical imagery. In my opinion, at least, the use of personal anatomy serves to intensify a given situation or dialogue. "You will not see *my face* again," for example, seems to carry more emotional impact than simply saying, "you will not see me again." Similarly, personal pronouns are often coupled with the "ear," which in turn, often accompanies the verbs, "hear," "say," "speak," etc., so as to intensify the imperative intent.

"Say in the ears of all the citizens of Shechem..." (Jgs 9:2); "...let your handmaid speak in your ears..." (1Sm 25:24); ...according to all we have heard with our ears." (2 Sm 7:22), are examples of this kind of usage. It is impossible to know exactly how the ancient Israelite may have responded to such (anatomical) imagery, but it seems to have been an important literary device among biblical authors. Wolff asserts that when an Israelite's hearing was in peril it was tantamount to threatening his very humanity. Above all, it is in the hearing, says Wolff, that makes a man—that and the ability to open the mouth and respond to what is heard (*Anthropology of the Old Testament*, 74-75). Wolf explains further, "Since human life is reasonable life, the hearing ear and the properly directed tongue are the essential organs of man."

It would be difficult to overestimate the importance of the faculty of hearing, especially to the ancient Israelite. One receives wisdom and is transformed by it through the organ of hearing. The very institution of a People of God begins with hearing the words of Yahweh. This relationship continues and is constantly reinforced in the lives of the people, such that, "...not by bread alone does man live, but by every word that comes forth from the mouth of the Lord" (Dt 8:3).

> Give Ear O heavens, while I speak;
> let the earth hearken to the words of
> my mouth.
> May my instruction soak in like the rain,
> and my discourse permeate like the dew,
> like a downpour upon the grass,
> like a shower upon the crops:
> For I will sing the Lord's renown.
> Oh, proclaim the greatness of our God!
> The Rock—how faultless are his deeds.
> how right all his ways,
> A faithful God without deceit,
> how just and upright he is.
> (*NAB*, Dt 32:1-4)

This beginning of the *Song of Moses* highlights the importance of hearing. The Hebrew word for hear (*shama*) often carries with it the implication to attend what is heard with attention and obedience. Not just the present, but all the future depends on hearing (and hearing again) these words. They will be remembered by a people, "a people living, sinning, and struggling in the dust of history" (Harold Fisch, *Poetry with a Purpose*, 64). The Song of Moses, says Fisch, echoes endlessly throughout the Psalms and the prophetic books, its poetry and words remain potent, in spite of the lapse of time. It will be laid up by the people (cf. Dt 32:40) to explode on them at some future need. "Heaven and earth" in the Hebrew imagination, asserts Fisch, "are significantly different from the Greek cosmos." They belong, rather, to the broader scheme of "a contractual history." When the contract is broken, which occurs frequently in the course of history, the prophet's oracles repeatedly cry out for Israel ***to hear***.

> Hear, O heavens, and listen, O earth,
> for the Lord speaks:
> Sons have I raised and reared,
> but they have disowned me!
> An ox knows its owner,
> and an ass, its master's manger;
> But Israel does not know,
> my people has not understood
> (*NAB*, Is 1:1-3)

As mentioned earlier, hearing plays an important role in defining what it is to be human. Thus, expressions such as, "say in the ears," or "incline your ear and listen," or "give ear, O heavens and hear, O earth," provide an appropriate emphasis to the importance of hearing, especially as it applies to human covenantal relationships with God. Indeed, hearing Yahweh is directly related to fearing him (Dt 13:11; 21:21; 31:12), and fear of the Lord, as the psalmist writes, "is the beginning of wisdom" (Ps 111:10). Even friendship with the Lord comes to those who fear him (Ps 25:14), as does his kindness and justice (Ps 103:17). Thus, all proper relationships with God begin with those who fear him, fear in this case implying a sense of reverence, rather than abject terror we might sometimes associate with this word (cf. Strong's Hebrew Dictionary, #8085).

In those ancient times it would have been difficult to imagine any way to reach a proper relationship with Yahweh in the absence of the ear that hears. As the deuteronomist writes:

> Assemble the people, men, women, and little ones, and the sojourner with your towns, that they may hear and learn to fear the Lord your God, and be careful to do all the words of the law... (Dt 31:12)

The importance and urgency of the cry to hear does not diminish in New Testament times. In the gospel of Luke those who hear and do the will of God become the mother and brothers of Jesus (Lk 8:21; cf. Mt 12:50; Mk 3:35). Paul asks the question (Rom 10:24), "how are [men] to believe in him of whom they have never heard? And how are they to hear without a preacher?" To Paul the very foundation of faith lies in hearing the gospel of Christ.

The Symbolic Ear: Why the Right One?

In the books of Exodus and Leviticus, sanctification of the whole man begins by touching the blood of a sacrificial animal to the right ear, then proceeding to the right thumb and then to the right large toe (Ex 29:20; Lv 8:24). Thus, the whole person is symbolically included in the religious ceremony, beginning with the right ear. It is important (for future reference), that the ritual begins with the right ear.

There is a possible New Testament connection to this practice, viz., as one of the disciples of Jesus draws his sword and cuts off the ear of the high priest's slave (Mt 26:51; Mk 14:47; Lk 22:50; Jn 18:10). The fact that this incident is reported in all four gospels increases the likelihood that it actually occurred.

The gospels of Luke and John go a step further and indicate it was the right ear which was severed. John alone identifies the sword wielder as Simon Peter. One has to question why the particular ear was preserved in the traditions of Luke and John. If it were simply a case of the disciple venting his frustration over the arrest of Jesus, why bother to report it and why, in particular, specify the right ear?

Another thought which occurs here is the difficulty of cutting off the right ear unless the assailant happened to be left-handed, or the priest's slave had his back to the disciple. Left-handedness was probably an uncommon trait among the Jewish people (see Ch. 7), and it would seem rather cowardly to sever an ear while the victim had his back to the one doing the deed. So why do the gospels of John and Luke make a point of identifying the particular ear of the high priest's slave?

The severed ear does heighten the drama of the scene, true enough, but does it really matter which ear was cut off? To me, the question begs the reader to look to some significance beyond the event itself.

To be sure, the prominence of the right ear as the first touched by the blood of the ram in priestly consecrations (Ex 29:20; Lv 8:24; cf. Lv 14:17), leads one to suspect that the authors of Luke and John, in particular, had a theological agenda in reporting the incident.

Attacking the slave of the high priest, and not another among the arresting body, indicates, in my opinion, that the high priest was the symbolic victim, rather than the one actually cut. Furthermore, cutting off the right ear was not simply an act of defiance, but a figurative invalidation of the priesthood itself. After all, it is this particular anatomical structure (the right ear) with which the priestly consecration begins. Removal of the structure upon which the beginning of the ritual depends seems highly significant, although the possibility of a coincidence remains.

Examining the narrative from this perspective brings new meaning to a seemingly unimportant incident. Certainly, against the background of Jesus' agony and impending arrest, the severed ear makes little sense, unless there is something here of theological importance to be gleaned.

Raymond Brown identifies two groups of Jews who may have constituted an early foundation for John's community. The first group may have been followers of John the Baptist, at least originally, and who had some relationship to the Essene Jews. Their "dualistic" views (light/darkness; truth/falsehood) shows some connection to the Essene community, those responsible for the "Dead Sea Scrolls" (*The Community of the Beloved Disciple*, p.30).

A second group, according to Brown (p. 38), were anti-Temple Jews who, in turn, probably converted Samaritan elements (cf. Jn 4:1-42). The Samaritan, anti-Temple components rendered John's community particularly obnoxious, says Brown, to traditional Jewish groups. It therefore seems that a considerable level of hostility existed between John's community and those Jews devoted to priesthood and Temple. In light of this environment of hostility, it would not be inconsistent with the symbolic invalidation of the priesthood by severing the right ear of the priest's slave.

I am reminded here of another seemingly unimportant incident, the cursing of the fig tree (Mk 11:12), which may have had symbolic meaning. The narrative makes no sense in isolation, since the fig tree was not supposed to bear fruit, "...for it was not the season for figs" (11:13). But the verses that follow tell of the cleansing of the temple. Like the fig tree, the temple (and barren Israel) was no longer bearing fruit (cf. Jer 8:13; 29:17). The gospel is saying, I believe, there is a more direct way, a more appropriate way of worship, and that is through Jesus the Christ.

The symbolic severing of the high priest's ear adds another dimension to the fig tree incident. What use does one have for priests when there is no longer a temple? Who needs a priest without burnt offerings? The fig tree narrative deposes any further need for a temple, followed by the symbolic invalidation of the priestly office. Both temple and the Jewish priesthood are irrelevant in the age of the new covenant.

One can only guess at why John identifies the sword wielder as Peter. Certainly, Peter becomes one of the most important of all the disciples in the early church. Perhaps John is attempting to show another relationship, or chain of command. Jesus came, a church was established in his name. Peter, acting in the church's behest, strikes the symbolic blow.

Luke is the only gospel that has Jesus healing the slave's severed ear. The author may have felt it necessary to show the inconsistency of such a violent act with the message of his gospel. Matthew's gospel brings about the same result by having Jesus demand of the attacker, "Put your sword back into its sheath, for all who take the sword will perish by the sword" Mt 26:52). The Jesus in John's gospel tells Peter to "Put your sword into its scabbard. Shall I not drink the cup that the Father gave me?" (Jn 18:11). In the gospel of Luke, Jesus gives none of these explanations for why such violence is inconsistent with his ministry. Jesus simply touches the injured slave's ear, and in a symbolic way brings order out of chaos, kindness out of retribution, compassion out of hatred. The healing of the slave's ear is a powerful gesture which, in my opinion, says much more than any words can possibly convey.

The Ears of God.

What kind of ears does God possess? This is a question that comes into the heart of the psalmist. It is not unlike Job's question to the Lord, "Do you have eyes of flesh? Do you see as human's see? (Jb 10:4). In a like fashion, the psalmists asks, "He who planted the ear, does he not hear? He formed the eye, does he not see? (Ps 94:9). These questions are a very simple way of exploring the idea of God's relationship to creation, particularly as regards human creatures. Beneath the surface of this innocent query is a much deeper presumption, perhaps tinged with a modicum of uncertainty. Indeed, *is God a being who cares what we humans say and do? Is their a reciprocal relationship between God and his human creatures? Did God put everything in motion and then sit back, uncaring, uninterested? What kind of God is God?*

> When I behold your heavens, the work of
> your fingers,
> the moon and the stars which you set in place—
> What is man that you should be mindful of him,
> or the son of man that you should care for him?
> (*NAB*, Ps 8:4-5).

The Greek philosophers, Aristotle and Plato in particular, may have couched such thoughts in different words, framed them in more sophisticated Greek idioms, but they were no less profound than the questioning of the ancient Hebrew people. To be sure, this is the eternal question—what kind of God is God—asked over and over and again in every generation, just as it will continue to be expressed for generations to come, as long as there are sentient beings to ask it.

Ears that Do Not Hear

> Make the heart of this people fat, and their ears heavy, and shut their eyes; lest they see with their eyes, and hear with their ears, and understand with their hearts, and turn and be healed. (Is 6:10)

This excerpt from Isaiah (above) is one of the most frequently quoted in the New Testament. It is referred to in all three synoptic gospels as Jesus' explanation of why he speaks in parables. This dialogue with the disciples follows the Parable of the Sower (Mt. 13:13-14;

Mk 4:12; Lk 8:10). In the fourth gospel, Jesus cites it as an explanation of the unbelief of the Jews, and as a fulfillment of Isaiah's prophecy (Jn 12:40), without reference to a parable. Similarly, the apostle Paul echoes the words of Isaiah as he speaks to the Jews from his prison cell in Rome (Acts 28:27). In his letter to the Romans, Paul expresses these sentiments in the context of "the remnant" of Israel, chosen by grace to believe. "The elect obtained it, but the rest were hardened" (Rom 11:5-8). In the verses that follow, Paul goes on to explain that God has not rejected his people. They (the Jews) have stumbled but not fallen, "but through their trespass salvation has come to the Gentiles" (Rom 11:11-12).

According to Paul, therefore, it seems that the fulfillment of Isaiah's prophecy is all a part of God's strategy for salvation of Gentile and Jew alike. I wish it were that simple, and perhaps it is, but why then is there such an argument among scholars as to the meaning of Isaiah 6:10? Why all the fuss? Why not simply accept the biblical texts as they stand and let it go at that?

One reason, I suppose, is that these passages seem to be saying that God is the agent of this hardening of the heart, this spiritual blindness and deafness. Paul tells us that God "has mercy upon whomever he wills, and he hardens the heart of whomever he wills" (Rom 9:18). Gerhard von Rad (*The Message of the Prophets*) makes no secret of his opinion on this issue (also see Chapter Two). He asserts that this is one means by which God becomes present in history to the people of Israel. Others reason that persistent rejection of God's will leads to mental and spiritual dullness. The more one preaches to ones of this inclination the more hardened of heart they become.

Regardless of how one interprets the words of Isaiah 6:10 (and similar biblical texts), most views coincide in the final analysis, i.e., that salvation history, God's strategy for us, is somehow involved. The striking aspect of this prophetic utterance is its distinctive combination of a fat heart (hard heart) with spiritual deafness and blindness. These three metaphors for spiritual obduracy seem to carry more weight when used together. Such an extreme state might be compared with total sensory deprivation, a condition (self imposed or otherwise), which can be the cause of bizarre hallucinations if not outright insanity. The difference between sensory and spiritual isolation, of course, lies in the consequences. According to Paul, living a life of sinful abandon eventually turns in on the perpetrator as God's wrath (Rom 2:5-6). God's wrath, then, appears to be something we do to ourselves, unlike the prototypal pagan god who may be characterized as throwing lightening bolts and playing tricks on mortal subjects.

Salvation, then, especially in the Christian sense, is being saved from our proclivity for sinful living. How that is done, biblically speaking, is a matter of God opening our eyes and ears to divine will, removing hearts of stone and giving us a new, spirit filled heart (Ez 36:26).

In worshiping idols of wood or metal, which can neither hear nor see, those who do so become like what they worship (Dt 28). This is the consequence of turning one's heart away from God (Dt 30:17). Thus, the persistent cry of the prophets is to incline the ear, to hear my voice, to give ear, etc., to what the Lord is telling the people (Is 28:23; 51:4), but, as often happens, they do not listen or incline their ear (Jer 7:26; 17:23; Ez 12:2). They have ears but do not hear and eyes that do not see.

But even in the presence of spiritual deafness and blindness the prophet is not relieved of his obligation to speak out. "Whether they hear or refuse to hear (for they are a rebellious house) they will know that a prophet is among them" (Ez 2:5; 2:7). It was the same problem for Isaiah (cf. Is 6:10), in that the prophetic calling was made all the more difficult because of the intransigence of the people. The more the prophet spoke the more they refused to listen. Jeremiah, growing weary of all this resistance, and the abuse he must suffer as a result, complains:

> You duped me, O lord, and I let myself be duped; you were too strong for me, and you triumphed. All the day I am an object of laughter; everyone mocks me. Wherever I speak, I must cry out, violence and outrage is my message; The word of the Lord has brought me derision and reproach all the day. I say to myself, I will not mention him, I will speak in his name no more. But then it becomes like fire burning in my heart, imprisoned in my bones; I grow weary holding it in, I cannot endure it... (*NAB*, Jer 20:7-9)

The tragic figure of Jeremiah continues to cry out because he is forced to do so. He cannot hold it in, even though he has concluded that his words are falling on deaf ears. As he says to the people of Jerusalem, "If you do not listen to this in your pride, I will weep in secret many tears; My eyes will run with tears for the Lord's flock, led away to exile" (Jer 13:17). He goes on to lament, "Can an Ethiopian change his skin? the leopard his spots? As easily would you be able to do good, accustomed to evil as you are" (Jer 13:23).

Hearing can center in on the beauty of something, the voice heard, for example, without attention to the spiritual nature of the message. This phenomenon becomes a particular problem of the prophet Ezekiel. The people would listen with their ears, hear the beauty of his words, but still not get the message. It is as though their ears were tuned to one wavelength and their hearts to another. The prophet's popularity was for the wrong reason, and in that sense, it was a false popularity. The people listened to the prophet's words as a musical entertainment, as an artistic performance.

As for you, son of man, your countrymen are talking about you along the walls and in the doorways of houses. They say to one another, "Come and hear the latest word that comes from the Lord." My people came to you as people always do; they sit down before you and hear your words, but they will not obey them, for lies are on their lips and their desires are fixed on dishonest gain. For them you are only a ballad singer, with a pleasant voice and a clever touch. They listen to your words, but they will not obey them. (NAB, Ez 33:30-32).

Graphic Symbolism of the Prophets.

Knowing they are speaking to a spiritually deaf people, to ears that do not hear, the prophets resort to all sorts of graphic symbolism in a desperate effort to get their message across. The Lord tells Jeremiah to go buy a loincloth, wear it for awhile, then, without washing it, hide it next to the river Pareth. After an interval, the Lord asked him to go and fetch the loincloth, but by this time "it was rotten and good for nothing!" Then the Lord said to Jeremiah:

> So also I will allow the pride of Judah to rot, the great pride of Jerusalem. This wicked people who refuse to obey my words, who walk in the stubbornness of their hearts, and follow strange gods to serve and adore them, shall be like this loincloth which is good for nothing (*NAB*, Jer 13:1-10)

One can imagine the shock and dismay of the people as Jeremiah held up the rotted loincloth before an assembled group, pointing to it and telling them this is what has become of the house of Israel. Similarly, one can envisage the outrage of the indolent women of Samaria, as the prophet Amos refers to them as fat cows (cows of

"Bashan," a very rich pasture land for fattening the herds). He identifies their offenses as "you who oppress the weak and abuse the needy; Who say to your lords, 'Bring drink for us!'" Then adds some of the most graphic imagery of the Bible, as he tells them:

> The Lord God has sworn by his holiness:
> Truly the days are coming upon you
> When they shall drag you away with hooks,
> the last of you with fishhooks;
> You shall go out through breached walls
> each in the most direct way,
> And you shall be cast in the mire,
> says the Lord
> (*NAB*, Am 4:1-3).

We have already seen how Jeremiah suffered in speaking the word of the Lord. He tells God he "was duped," and all he got from being a prophet was abuse. People even laughed at him. Amos had no ambition to be a prophet. He tells Amaziah, the priest of Bethal, "I was no prophet, nor have I belonged to a company of prophets; I was a shepherd and a dresser of sycamores. The Lord took me from following the flock, and said to me, 'Go prophesy to my people Israel'" (Am 7:14-15).

One can imagine how difficult the life of a prophet must have been. They had to point out the sins of the people, something they did not want to hear in the first place. If they listened at all, they had ears that did not hear. They were blinded by selfish and prideful living, taking offense at everything the prophet said. "To whom shall I speak and give warning that they may hear," complains Jeremiah, "Behold, their ears are closed, they cannot listen; behold, the word of the Lord is to them an object of scorn. . ."(Jer 6:10).

In desperation, perhaps, or for divine prerogative not made known to us, the prophet frequently resorted to language that further alienated him from the people. Telling them they were like rotten underwear, or that they resembled fat cows, did little to endear the prophet to his audience. These men were the real tragic figures of the Bible, and at the same time, its true heroes. Except for what comfort they took in their intimacy with the Lord, they must have been lonely beings, indeed. But they were obedient to the point of seeming absurdity, even as Ezekiel was commanded (in a vision) to prophesy to a plain of dry bones (Ez 37:1-14).

Why Bother to Prophesy?

So why go to all the trouble? Why bother to prophesy when nobody, or almost nobody, will listen to the words? Even if they listened, they did so with ears that were closed. A partial answer comes to Ezekiel when the Lord tells him, "whether they hear or refuse to hear...they will know there has been a prophet among them" (Ez 2:5). The symbolic prophecy to dry bones (representing Israel), would lend future hope, through the image of resurrection, to an otherwise desperate group of exiles (Ez 37:12).

Of course, the people of modern times presume to know another reason why these ancient words were spoken—for future generations. Although there are among us today many who do not hear, or refuse to listen, there are millions of us that do hear, avidly, and are comforted with the hope of expectation. To be sure, the inspired words of the prophets will continue to ring out their message of justice and comfort. Those who suffer, all those weak and oppressed numbers among us, will endure because of those words of hope. The prophetic voices from those ancient times will always be relevant, no matter how far humankind ventures into the age of science and technology.

Difficult Biblical Imagery.

Some of the imagery of the Bible is difficult to understand from our modern perspective. This often difficult, figurative language, is one of the major reasons modern translations of the Bible are necessary. In the chapter on the heart I mentioned the image of the "uncircumcised heart," an often used metaphor that means the same thing as being hard hearted, not open to the word of God. This particular imagery hearkens back to the relationship of circumcision and covenant. The picture forced upon us here is that of a covering, not unlike a penile foreskin, which closes off the heart, thus blocking receptiveness to the will of God.

The "uncircumcised ear" is mentioned in the Acts of the Apostles (Acts 7:51), along with the afflictions of being "stiff-necked" and of "uncircumcised heart," all of which amount to the same thing. The apostle Paul asserts that "He is a Jew who is one inwardly, and real circumcision is a matter of the heart, spiritual and not literal" (Rom 2:29). One would suppose that an uncircumcised ear, like the heart, is in need of metaphorical surgery in order to relieve the afflicted person of his/her spiritual deafness. If this is so, one could safely paraphrase the deuteronomist saying, "Circumcise, therefore, the foreskin of your [ears], and be no longer stubborn" (cf. Dt 10:16).

There is also such a thing as "uncircumcised lips" (Ex 6:12), which is a literal translation of the Hebrew Bible. Some versions translate Moses' affliction as "poor speaker" (*NAB*); or "slow of speech" (*JB*). To my knowledge this is the one case of "uncircumcision" that does not imply hard heartedness (see Chapter Six for further discussion). The covering of the lips (metaphorical foreskin) does not obstruct openness to God's will as do the others. Moses' affliction, rather, is more akin to the "unclean lips" of Isaiah. According to Paul, even uncircumcision of the penis is not an affront to God (Rom 2:29). Uncircumcised ears and hearts, however, remain a serious spiritual affliction in Old and New Testaments.

The Seriousness of Spiritual Deafness.

The seriousness of spiritual deafness in the New Testament is brought home in the touching and tragic parable of "The Rich Man and Lazarus" (Lk 16:19ff). When the rich man dies and finds himself in Hades it is too late for repentance. He looks across the "great chasm" dividing him from the bosom of Abraham and sees Lazarus being comforted there. He begs for mercy to no avail, because the chasm is fixed, across which none may pass. The rich man then begs Abraham to send someone to his brothers "that he may warn them lest they all come into this place of torment." Abraham responded, "they have Moses and the prophets; let them hear them." But the rich man, knowing his brothers were spiritually deaf and blind like himself while he lived, said, "No, father Abraham; but if someone goes to them from the dead, they will repent." Then, in a most dramatic and poignant response, Abraham says to the rich man, "If they do not hear Moses and the prophets, neither will they be convinced if some one should rise from the dead."

CHAPTER SIX

The Mouth

> *But Moses said to the Lord, "Oh, my Lord, I am not eloquent, either heretofore or since thou hast spoken to thy servant; but I am slow of speech and tongue." Then the Lord said to him, "Who has made man's mouth? Who makes him dumb, or deaf, or seeing, or blind? Is it not I, the Lord? Now therefore go, and I will be with your mouth and teach you what you shall speak."* (Ex 4:10-12)

I have a friend who is a nutritional biochemist. In his research he studies the body's normal needs as well as the nutritional requirements of the body under stress. One interesting finding he told me about was the body's need of the element zinc, especially under stress. Following a severe injury or some kind of extensive surgical procedure, quantities of zinc pour out of the body through the urinary tract. Furthermore, an increase in dietary zinc following a severe trauma accelerates the healing process.

Hearing of my friends research (some years ago) brings to mind the close association of the body's nutritional needs with the whole idea of "spiritual nutrition." Indeed the two are so intimately linked it is hard to say where one ends and the other begins. When we are ill, accidentally injured, or hospitalized for any reason, where do we draw the line between our physical needs at the time and our need for spiritual comfort and reassurance? We are always in need of spiritual nourishment, but just as our body requires special dietary care after it is severely traumatized, so does our spiritual side. One simply cannot say, right now all I need is food, to hell with anything else. Studies have shown repeatedly that the hospitalized person who has a strong relationship with God recovers faster and more completely. In other words, there is a spiritual zinc that comes to our aid in times of stress that augments the required physical nourishment. Without question the two go hand in hand.

The biblical writers knew of this relationship between body and spirit, and made use of it. In doing so, the inspired authors of the Bible were able to express profound religious truths in very simple terms. They were able to use the mouth, for example, as a symbol for so much more than the simple eating, drinking and speaking. To be sure, they were very skillful in their use of this figurative sense, and used it to illustrate the deeper significance of a life lived in relationship to God. The interplay of the mouth that eats with the nourishment of the human spirit, therefore, demonstrates a profound sophistication in the thinking of the Hebrew authors of the Bible

That is what this chapter is all about. The Greeks had their metaphysical techniques for exploring abstract ideas, just as the biblical authors had their own tools for uncovering truth. The biblical authors were no less profound in their search for ultimate truth. They just did so in a more easily understandable context, that is, the human body and its basic functions.

Here, the term "mouth" is intended to include associated structures, such as tongue, throat, palate, etc., or any oral anatomical parts used in eating, drinking, or speaking. Hans Walter Wolff (*Anthropology of the Old Testament*), makes it clear that ancient Hebrew uses "one and the same word" to express different ideas, whereas the English language may use widely differing ones. In each case it is the context of the biblical verse that determines the meaning of a given Hebrew word. In the case of oral structures, in order to avoid the highly complex, hair-splitting translations of the Hebrew word, I will use the fairly literal, *Revised Standard Version*, as a guide. Otherwise, it would be very difficult to reach conclusions concerning various meanings of the Hebrew word or their applications. Our goal here is not a precise definition of Hebrew semantics, but a grasp of the religious and social implications of, in this case, the word, "mouth," and its function in eating, drinking, and speaking. The task becomes much easier in the Greek New Testament, since here the word for mouth (*stoma*) is more consistent with the modern understanding.

In English we may use the word "mouth" for any opening or orifice, such as the mouth of a cave, a river, a sack, or other inanimate objects. The point I wish to make here is, although the Hebrew might use a different word where we tend to use the same one, the translation will usually be, "mouth." This makes our analysis simpler, because we can take advantage of the analogies made between an inanimate and living "mouth" without confusing the issue with different words.

Like the "face," the word "mouth" may be applied to inanimate objects as well as to the human opening used for speaking and eating. The "mouth" of a cave, a sack, a river, and so on, are commonly found usages in the Bible as well as in various modern languages. The Old Testament, however, expands the analogy of the human mouth with inanimate objects beyond that commonly found in modern English. The image of the mouth as a container similar to a sack, for example, is a particularly useful analogy. Even if one is not busily eating at the time, the mouth is still a container of sorts. In addition to food, the mouth is a container for words, just as a sack may hold grain or some other useful material (Nm 23:5; Dt 18:18; 2 Sm 14:3; 14:19). In instances where the Lord is the one who places words in the mouth, the act carries with it an obligation to keep them, or lay them to heart, so that future generations may also profit from them. The call of Jeremiah (Jer 1:9), is an example of this obligation to carry the words of the Lord for future prophetic utterances. The quote below from 2nd Isaiah also makes this mandate abundantly clear.

> As for me, this is my covenant with them, says the Lord; my spirit which is upon you , and my word which I have put in your mouth, shall not depart out of your mouth, or out of the mouth of your children, or out of the mouth of your children's children, says the Lord, from this time forth and for evermore. (Is 50:21)

Thus, the idea of the mouth as a container of words carries with it the imagery of something which can be filled or emptied as the given situation requires. The difference between a filled sack and the mouth full of words, is that the mouth is never really emptied.

Words can also be eaten, as an extension of this kind of imagery (Ez 2:8; 3:2) and can even be tasted (Ez 3:3; Ps 119: 103; Rv 10:9). Indeed, words can even be thought of in terms of a fruit of the mouth (Prv 18:20), especially if they are words of wisdom. Thus, words are like seeds that can bear fruit. The fruit of the mouth is not always good, however, since lies can also bear fruit (Hos 10:13).

It is this kind of language that makes the Bible so rich in its imagery. It is one reason why literal translations can be more rewarding, if only one takes the time to understand how the ancient writers used body parts to illustrate their thinking. The Hebrew writers knew nothing of how the brain receives information, or stores it, such that it can later be retrieved in some form close to the original utterance. For the ancients, it was easier to think in terms of the mouth as a carrier of words which, in turn, could be released from their container as required.

Perhaps, we have retained the rudiments of this kind of thinking. Putting a hand over the mouth, for example, is a gesture commonly indicating a desire not to speak out improperly. It is a mannerism which implies holding words prisoner in the mouth. As with many of our commonly used mannerisms and expressions, some may hearken back to ancient times, when references to body parts were common, and usually implied very complex thinking.

The Figurative Mouth

The mouth becomes a useful tool for metaphorical expressions, as well as other kinds of figurative language. There are many examples of this kind of figurative language in the Bible, which is why it is such a richly expressive literature. Rather than attempt a clear and concise definition for these kinds of expressions, the reader might be better served with examples from the Bible.

The Mouth as a Weapon.

The mouth can be a metaphor for a weapon, for example, such that the source of words emanating from it can be the cause of violence in a given situation. In Hosea (Hos 6:5), the weapon is implied, whereas the metaphor for mouth/weapon is designated as a rod in Isaiah (11:4), a sword (49:2), or some other destructive agent. Some of these metaphorical weapons refer to the mouth of the prophet himself, whereas others carry with it a definite messianic ring (Is 11:4).

The words from the prophet's mouth are often just as unwelcomed as any weapon might be, perhaps even more so. As shown earlier, unpopularity among the people, even hatred of the prophet, was a common experience of those chosen by God to speak out in his name. The mouth, or its content of words as a metaphor for a weapon, are aptly expressed by Isaiah: "He made my mouth like a sharp sword, in the shadow of his hand he hid me; he made me a polished arrow, in his quiver he hid me away" (Is 49:2). In a similar vein the Lord informs Jeremiah, "I am making my words in your mouth a fire, and this people wood, and the fire shall devour them" (Jer 5:14). Surely the people would welcome a weapon they could see, more than one composed of words. Certainly the people did not want to hear Isaiah tell them of the mouth of Sheol about to devour them. It has "enlarged its appetite" for the nobility of Jerusalem as well as for the multitudes (Is 5:14). The image of Sheol, the place of the dead, being engorged with the people of Jerusalem was no happy scene for the people to contemplate. It is possible to defend oneself against a visible sword or an arrow, but a weapon of words may come without warning, reminding the victim of things and possibilities not welcome in the realm of conscious thinking.

The Mouth of God.

Another common literary device concerning the mouth is its use in connection with God. Such usage, in which a human feature is attributed to God, is referred to as an "anthropomorphism." It is obvious to us, if we believe that God is a spiritual being, that "he" does not possess a human mouth, eye, ear, or any other anatomical attribute. In this sense, even the use of "he" or "she" is an anthropomorphism, since as far as we know, God does not have a sexual gender. Nevertheless, even in modern parlance, we tend to attribute human characteristics to God. It makes us feel closer to God, perhaps, or at least makes such an awesome being more approachable. When Job asks God does he have eyes of flesh (Jb 10:4; cf. Chapter Four), the question is really much more profound than the literal meaning might imply. In so many words, Job is asking "what kind of God are you, God?" In the same way, our attributing to God certain human attributes is an attempt to make God more understandable, or, as mentioned earlier, more approachable.

The Mouth as Synecdoche.

Thus God, along with other human attributes, is said to have a mouth (Dt 8:3; Is 62:2; Jer 9:20; 23:16; Ps 138: 4). These references are just a few of many, since "the mouth of God" is a fairly common phrase. We might expect this, since God's words, a synonym for divine revelation, come from his "mouth." One would predict God to have a "mouth" in the thinking of the Hebrew writer, because it would be natural to associate divine wisdom with a source familiar to the ordinary person. Thus, the "mouth" of God.

The "mouth of God," like the hand, finger, or foot of God, can also be construed as representing "all of God," a literary device known as "synecdoche." The same is true with human creatures, whereby a part of the body can represent the whole person. "And now your eyes see, and the eyes of my brother Benjamin see, that it is my mouth that speaks to you" (Gn 45:12). Here, although there is an emphasis on the mouth speaking, the sense of the verse is that the entire person, in this case Joseph, is in the presence of his brothers. Thus the mouth of Joseph represents all of Joseph. Other translations (cf. *NAB* 45:12), may dispense with literal sense and insert, "it is I, Joseph," for "mouth." The question here becomes, is the reader missing some subtlety of meaning in not rendering the more literal sense of the biblical verse? To me, the next best thing to reading the Bible in the original Hebrew or Greek, is to read and understand the most literal of translations. In this way one is less likely to miss out on some of the more important nuances of the original language.

In a quite different sense, the mouth can represent the whole person in terms of his/her needs (cf. Wolff, ch. II). Physical needs of the individual, for example, can be reckoned in terms of the food that sustains the body. By contrast, there are other needs, not excluding bodily ones, that relate to the whole person. Qoheleth laments that, "All the toil of man is for his mouth, yet his appetite is not satisfied" (Eccl 6:7). One can find an explanation for this difficulty in terms of another mouth, i.e., the "mouth of God." In the words of the deuteronomist, "man does not live by bread alone, but by everything that proceeds out of the mouth of the Lord" (Dt 8:3). Here the mouth that eats food and the mouth of God are related to one another with respect to the needs of the whole person. There are other ways, as I hope to make clear in the next section, that the human mouth is important to the physical and spiritual needs of the individual and to the community as well.

The Mouth that Eats

> Now as they were eating, Jesus took bread, and blessed, and broke it, and gave it to the disciples and said, "take, eat; this is my body." (Mat 26:26)

The *mouth that eats* is one of the most unifying themes in the entire Bible. Indeed, theological threads related to eating weave their way from the book of Genesis to the New Testament, linked together by a consistent relationship to the idea of redemption. In what follows I have identified certain events involving the food of the Bible with redemptive history. Some may think of better examples, but I believe the case can be made easily from the events I have chosen to illustrate this relationship. Considered in the context of saving history, the food of redemption consists of a spiritual reality which opens the eyes of the blind and the ears of the deaf, and nourishes the heart with its intrinsic beauty.

Beginning with the second creation story and continuing throughout the New Testament, food and the eating of it are central to the story of redemption. In the garden of Eden there was one tree, that of "Knowledge of Good and Evil," which was forbidden to the first humans (Gn 2:9). But the woman, tempted by the serpent, saw that the fruit of this tree was "a delight to the eyes, and that the tree was to be desired to make one wise" (Gn 3:6). So she took the fruit of the forbidden tree and ate, also giving some of it to her husband.

This simple story contains, embedded within it, a very profound truth about the nature of humankind. One does not have to accept the literal reading of the story of Adam and Eve in order to retrieve from it the insightful understanding of human nature it contains. It could be said, had there been no temptation, no test of human freedom as revealed in this simple story, there would have been no need for a redemptive history. God, however, created us as free human beings, and from that moment, we became our own worst enemy. We could have been created as automatons, as glorified biochemical robots, in which case redemption would have been an absurdity. Instead, scholars tell us we were created **from love** with a capacity **for love**. But one cannot love unless there is a **choice not to love**. Therein lies the crux of redemptive history.

Savory Food.

Jacob, who becomes "Israel," is of key importance to the story of redemption. Central to this story is the "savory food" which the older son, Esau, is to prepare and bring to his father, Isaac, in order to receive the father's blessing. Isaac's wife, Rebekah, overheard the discussion and informed Jacob of it. She advised him to go to the flock and bring her "two good kids, that I may prepare from them savory food for your father, such as he loves" (Gn 27:9). Jacob did as Rebekah instructed him, thus deceiving the father into believing he was Esau. Because of Isaac's poor sight, the charade was successful, and Jacob received the father's blessing instead of his older brother, Esau (Gn 27:18-29).

Jacob becomes "Israel," of course, and his twelve sons will provide the seeds of the tribes of Israel. One might ask, where would the people of God come from if not for the blessing of Jacob. Jacob's deception is neither approved nor condemned in the course of the story. It is, like so many biblical events, illustrative of how God's strategy for human redemption works within the framework of human weakness. The Hebrew author is illustrating by means of ordinary things, such as the "savory food" Jacob serves his father, how God is present to his people in history.

The Great Famine.

The role of food, or lack of it (famine) is crucial to the unfolding of this important story in the history of redemption. Joseph, one of the twelve sons of Jacob (Israel) is sold into slavery by his jealous brothers (Gn 37:1-36). Joseph appears initially as a spoiled son, favored by his doting father. Jacob adds to the distressing atmosphere of the family by giving Joseph his favorite piece of clothing, an item that marks his elevated status in the family. Adding insult to injury, Jacob freed Joseph from work among the flocks, a burden which had to be carried on by the other sons. The brothers succumb to sibling jealousy and plot an occasion to kill Joseph. Their resolve increases when Joseph reports dreams that confirm his pre-eminence. Therefore, when Joseph leaves the protection of his father in order to seek out the brothers' whereabouts, the brothers make their move.Their initial plan to kill Joseph, however, changes to a more economically advantageous one, that is, to sell Joseph to a passing caravan of merchants. The alternative plan is tantamount to the killing of Joseph, but at least the brothers have no blood on their hands.

The merchants sell Joseph into slavery in Egypt, but happily, Joseph rises from his status as a servant to become a wise and prudent administrator in Pharaoh's court. Joseph's good fortune came about because he correctly interpreted Pharaoh's dream, which predicted a great famine (Gn 41). With Joseph's help, Pharaoh was able to prepare for the disaster, thus saving the people of Egypt from starvation. Elevated to second in command over the entire kingdom, Joseph received responsibility from Pharaoh to administer the grain reserves in the time of famine. He did this for the benefit of the Egyptians and for the surrounding nations as well.

As irony would have it, the famine forces Joseph's brothers to travel to Egypt to buy enough grain to sustain their families (Gn 42-43). Once there, they stand at the mercy of Joseph, whom they do not recognize. Joseph, on the other hand, immediately recognizes his brothers, and in a dramatic series of exchanges, Joseph finally reveals his true identity (Gn 45).

Reconciliation between the brothers occurs, and Jacob (Israel), with his families in Canaan, migrates to Egypt. The reunion brings peace to a family once at war within itself. This is the biblical story of how the Israelites came to live in Egypt from which they would later have to be delivered. Their favorable status changed, probably with the overthrow of the "foreign rulers," the Hyksos Pharaohs. This change in status for the Israelites is reflected in the first chapter of Exodus (Ex 1:8).

A story of "savory food," followed by the absence of food in the form of "a great famine," have led the people of God into Egypt. Starting with the book of Exodus, their favorable condition changes for the worse. As slaves in Egypt, the stage is set, biblically speaking, for another crucial event involving food, the *Passover feast*.

The Passover Meal.

The next important account of food and its eating is the story of the Passover meal, one of the greatest of all redemptive symbols. This occurrence precedes the exodus of the people from slavery in Egypt, bringing them ever closer to another key symbol, the Land of Promise, the "land of milk and honey." But they must first wander in the desert for forty years, undergoing many hardships and privations, including hunger and thirst. At times the people long to return to slavery and the flesh pots of Egypt, where at least they had plenty to eat. This great journey begins as the Lord prepares the people for the very special feast of the Passover.

> The Lord said to Moses and Aaron in the land of Egypt...tell all the congregation of Israel that on the tenth day of this month they shall take every man a lamb according to their fathers' houses, a lamb for a household; and if the household is too small for a lamb, then a man and his neighbor next to his house shall take according to the number of persons; according to what each can eat...when the whole assembly of the congregation of Israel shall kill their lambs in the evening. Then they shall take some of the blood, and put it on the two doorposts and the lintel of the houses in which they eat them. They shall eat the flesh that night, roasted; with unleavened bread and bitter herbs they shall eat it...In this manner shall you eat it: your loins girded, your sandals on your feet, and your staff in your hand; and

> you shall eat it in haste. It is the Lord's passover. For I will pass through the land of Egypt that night, and I will smite all the first-born in the land of Egypt, both man and beast; and on all the gods of Egypt I will execute Judgment: I am the Lord. The blood shall be a sign for you, upon the houses where you are; and when I see the blood, I will pass over you, and no plague shall fall upon you to destroy you when I smite the land of Egypt. This day shall be for you a memorial day, and you shall keep it as a feast to the Lord; throughout your generations you shall observe it as an ordinance for ever (Ex 12:1-14).

This ancient event becomes an important ritual for Christians. It is a reminder that Christ freed the people from the slavery of sin, just as Moses freed the people from the physical reality of slavery in Egypt. Following the exodus from Egypt, the wandering of the people in the desert becomes a symbol of one's journey through life. In this journey the faithful struggle against the temptation to return to a life of slavery (the slavery of sin). Enduring these trials ultimately brings the faithful to the Land of Promise, the land of "milk and honey" (Ex 3:8; Lv 20:24; Nm 13:27; Dt 6:3). In this new land there will be other temptations, symbolized not by lack of food and endless wandering in the desert, but by its "milk and honey."

> For when I have brought them into the land flowing with milk and honey, which I promised on oath to their ancestors, and they have eaten their fill and grown fat, they will turn to other gods and serve them, despising me and breaking my covenenant. (*NRSV*, Dt 31: 20)

It is as though the "milk and honey" served as a catalyst for the people to bring more sins upon their heads. They did not have to worry so much about food and drink, as they did in the desert, but this only gave them time to invent other ways of sinfulness (cf. Jer 32:22-23).

The geographical Land of Promise, Canaan, symbolizes for Christians a place beyond physical reality, the divine dimension of God, the kingdom of heaven. To be sure, many of the historical events of the Old Testament become spiritually significant in the everyday lives of the Christian faithful. The "milk and honey" may be likened, in this case, to God's grace and faithfulness in bringing about salvation.

The Ultimate Food of Salvation.

For Christians the climactic moment of redemptive history arrives as Jesus becomes the lamb of God in his sacrificial death on the cross. This Christ-event has been made present in the lives of believing Christians ever since. Indeed, it becomes present again at the Eucharistic celebration, in which the faithful receive the body and blood of Christ in the form of bread and wine. Each time we open our mouths to receive the Eucharist, God becomes present to us in a special way. We faithful are asked to "do this" in remembrance (Greek, *anamnesis*) of Christ. We may also be reminded of the long history of God's presence in the lives of our ancestors, all the way back to the beginning of human existence. In this way food, and the eating of it, play a special role in our spiritual lives as well as in our physical well-being (see "The Mouth that Drinks," for further discussion of the Eucharistic celebration).

Foods in Other Words.

Aside from the relationship of food to the passover meal and the Christian Eucharist, there is a long tradition of food being used as a religious offering (Lv 3:11; 3:16; Nm 15:19; 28:2; Ez 42:13). There is a cereal offering (Lv 6:7ff), a sin offering (Lv 6:17ff), guilt offering (Lv 7:1ff), and the offering of peace (Lv 7:11ff). Unclean foods are important, probably as a practical matter as well as for religious reasons. The unclean foods listed in chapter eleven of Leviticus include the camel, the pig, hoofed animals that are not cloven footed or do not chew the cud (the horse and the ass). Also unclean are creatures which crawl or swim in the water but lack fins or scales, various insects, birds, creatures that swarm on the ground, and so on. It is these kinds of detailed proscriptions against certain foods, among other things, that make the book of Leviticus so tedious to read.

Food, and the eating of it, is such an integral part of everyday life that it is not surprising that it should become so central to the practice of religion. The multiplication of loaves, a miracle which first appears in the Old Testament (2 Kgs 4:42-43), occurs in all four gospels. In every instance of this miracle, a large number of people are fed from a seemingly inadequate source of food. Divine intervention plays an important role in this not so subtle reminder that the Lord provides for his human creatures. Food (and drink) even plays a role in the genealogy of Jesus, for after Boaz had eaten and was merry, it was then that Ruth made her move (Ru 3:7). Water to drink is a key to the identification of Rebekah, who becomes the wife of Isaac (Gn

24:14).

Just as the eating of food can be of religious significance, doing without it, or fasting, is usually construed as a means of worship. Doing without food implies a degree of obeisance and prayerful sacrifice. It is an intentional offering which, in a certain sense, is a gift of self that requires some level of suffering. John the Baptist, living the austere life of a holy man, dieted on locusts and wild honey (Mt 3:4). Christ went into the desert and fasted forty days and nights, overcoming temptation by the devil (Mt 4:1ff).

Other foods of religious significance are not food at all, at least not in the sense of receiving bodily nourishment. Eating of certain foods becomes a metaphor for spiritual unity and communication with the Lord. Ezekiel's eating of the scroll, for example, graphically illustrates the prophet's call to action (Ez 2:1-9). Food can also be used to symbolize the spiritual condition of the people, rulers, etc., as in the revelation of the two baskets of figs which the Lord showed Jeremiah (Jer 24:1-10). Jeremiah and Ezekiel alike use the metaphor of sour grapes to illustrate the importance and reality of personal responsibility (Jer 31:29-30; Ez 18: 1-32). The proverb against which the two prophets protest tells of fathers who have eaten sour grapes putting their children's teeth on edge.

> As I live, says the Lord God: I swear that there shall no longer be anyone among you who will repeat this proverb in Israel. For all lives are mine; the life of the father is like the life of the son, both are mine; only the one who sins shall die (*NAB*, Ez 18:3-4).

Proverbs (1:31-33) makes it clear that "fruit" must be eaten in it own way. It is clear from the context that the food, in this case, fruit, is a metaphorical expression for personal behavior. Other proverbs may be viewed in the same way (Prv 23:3; 30:8). As in the case of all metaphors of this type, they can be somewhat ambiguous, requiring thought and reflection to fathom their full meaning. Food is usually taken to mean that which nourishes the flesh, the biological necessities of sustaining life. When food is used in a metaphorical sense the meaning becomes clear in proportion to the one's wisdom and willingness to think beyond mere necessities. Thus, when the prophet Micah says, "You shall eat, without being satisfied, food that will leave you empty" (Mi 6:14), he may be talking about real food, but the impact of the words implies another kind of nourishment. Here, the prophet seems to refer to a food that is absent from the human menu, a metaphorical food to be sure. This lack of satisfaction goes far beyond mere food, yet points to something necessary for the sustenance of the whole person. Thus the human body requires more than any tangible food could ever provide. It needs spiritual nourishment as well.

The Mouth that Drinks

> But whoever drinks of the water that I shall give him will never thirst; the water that I shall give will become in him a spring of water welling up to eternal life. (Jn 4:14)

There is a considerable overlap, as one might expect, between "the mouth that eats" and the "mouth that drinks." Having said that, the act of drinking, in either the physiological or symbolic sense, has special significance. Jesus does not say anything about eating to James and John, for example, in reference of his suffering and death. He asks them, rather, "Are you able to drink the cup that I am to drink?" (Mt 20:22; Mk 10:38; cf. Jn 18:11). Similarly, at Gethsemane Jesus says to his Father, "if this cannot pass unless I drink it, thy will be done" (Mt 26:42).

In many other instances, eating and drinking are brought together in the same or adjacent verses. There are many references to drink and food offerings, particularly in the books of Leviticus and Numbers. Speaking of what is to pass following "the cup" that Jesus must drink, he tells his disciples that "you may eat and drink at my table in my kingdom..." (Lk 22:30).

The Last Supper.

The most celebrated of eating and drinking events, of course, is the tradition of the (Lord's) last supper:

> For I received from the Lord what I also delivered to you, that the Lord Jesus on the night when he was betrayed took bread, and when he had given thanks, he broke it, and said, "this is my body which is for you. Do this in remembrance of me." In the same way also the cup, after supper, saying, "This cup is the new covenant in my blood. Do this, as often as you drink it, in remembrance of me." For as often as you eat this bread and drink the cup, you proclaim the Lord's

death until he comes (1 Cor 11:23-26).

In the gospel accounts of the last supper (Mt 26:26-30; Mk 14:22-26; Lk 22:17-20), there are minor variations from that of the apostle Paul. In the gospel of John, however, a focal event of the last supper is the washing of the disciples' feet, whereas the Eucharistic formula found in Paul and the synoptic gospels is not mentioned (Jn 13:1-5). Allusions to the Eucharistic sacrifice, however, are brought out in other places in the fourth gospel (Jn 6:53-56).

In John's gospel, the idea of immanence, or the indwelling of God, is closely related to the receiving of the Eucharist, as is the idea of eternal life. "He who eats my flesh and drinks my blood has eternal life, and I will raise him up on the last day" (Jn 4:14). Jesus then goes on to add, "he who eats my flesh and drinks my blood abides in me, and I in him" (Jn 6:55). These words reinforce what Jesus said to the Samaritan woman (4:14), as he revealed to her the "living water" he gives, and its relationship to eternal life. Further on in the gospel Jesus tells the people that, "rivers of living water" will flow within those who believe in him (Jn 7:38). These events make eating and drinking a focal center of John's gospel, especially as they relate to the last days and the promise of eternal life. It prepares the way for what many refer to in the fourth gospel as "realized eschatology." As a case in point, Jesus said to Lazarus's sister, Martha, "I am the resurrection and the life; whoever believes in me, even if he dies, will live, and everyone who lives and believes in me will never die" (Jn 11:25-26).

The idea of thirst and water, representing spiritual thirst, and what will satisfy that thirst, the water of the Spirit abiding in us, cannot be isolated, or considered as somehow separate, from the Eucharistic celebration. It can be safely said that much of what Christians believe concerning the Eucharist looks to John's gospel for proof texts.

John's presentation of God's indwelling, as he depicts in the "abides in me and I in him" statement (6:56), closely parallels the meaning of Paul's "temple of the Holy Spirit" (1 Cor 6:19). In both cases, it is the idea of God's own Spirit living within us that gives such encouragement and fortitude to the Christian believer. It is this sense of immanence, the indwelling God, that allows the believer to think of God as a friend, and not simply a remote, uncaring deity. John's metaphorical presentation of thirst and water make this simple truth understandable to the average person.

It is these kinds of textual messages in John, as in the letters of

Paul, that prompted the Greek fathers to the insightful conclusion *that God became human so that humans could become like God.* The dominant strain in Athanasius' ideas on salvation, writes J.N.D. Kelly (*Early Christian Doctrines*), "is the physical theory that Christ, by becoming man [including his redemptive crucifixion] restored the divine image in us" (cf. Fitzmyer, *Pauline Theology*). The whole idea of "divinization" (Greek, *theopoiesis*), would seem absurd in the absence of John's high Christology and the imagery of "living water." Add to that Paul's ideas on the human body as the "temple of the Holy Spirit," or the faithful Christian viewed as a "new creation," (2 Cor 5:17; Gal 6:15), and the conclusion is inescapable: **we are being transformed** (cf. 2 Cor 3:18). This transformation is all because, "we are God's children now; it does not yet appear what we shall be, but we know that when he appears we shall be like him, for we shall see him as he is"(1 Jn 3:2).

Figurative Drinking.

Drinking in the Old Testament is often couched in figurative terms. The book of Job speaks of one who "drinks iniquity" (Jb 15:16). In the Psalms one can drink "from the river of delights" (36:8), or have "tears" to drink (80:5). In Proverbs one can drink the "wine of violence" (4:17; 26:6). The prophet Jeremiah speaks of drinking God's wrath (25:15), while Isaiah tells of no longer having to drink of it (Is 51:22). This kind of figurative language may not seem important if taken in isolation. The Hebrew writers, however, prepared the way for the more sophisticated metaphors like "living water," or the "cup" that Jesus must drink. It is in and through this kind of figurative language that the Hebrews were the equal of the Greeks, even superior to them, in expressing complex theological ideas.

The Mouth that Speaks

> Death and life are in the power of the tongue, and those who love it will eat its fruits (Prv 18:21)

This will be a brief section compared to "The Mouth that Eats," since much of the material is covered under previous headings. For example, it would be difficult to separate "the mouth that speaks," especially of wisdom, from "the ear that hears." Nevertheless, there are certain topics that merit separate inclusion under this heading. Much of the act of speaking in the Bible is expressed in terms of the "lips" and the "tongue." One reads of "lying," and "truthful lips," as well as a "smooth," "perverse," "mischievous," or a "gentle tongue," all in the context of speaking (cf. Is 35:6; 50:4; 59:3; Ps 66:14; 89:34; 119:13). In addition, one can speak metaphorically from the heart as did Hannah (1 Sm 1:13), although her lips were moving. Earlier, it was pointed out that the mouth could be considered as a container for words. In that context the lips can be construed as the purse strings, since words proceed out of, or pass through the lips (Nm 23: 23; 30:12). By contrast if the lips are closed, or guarded, nothing can escape them. And like the mouth, or almost any body part for that matter, the lips and tongue can be used to represent the entire person (synecdoche). This is a common literary device, especially in the Old Testament (cf. Jb 13:6; 27:4).

The uncircumcised lips of Moses (Ex 6:12; 6: 30) have been mentioned in connection with the inability to speak eloquently. To be sure, all but the most literal of translations will not mention "uncircumcised lips." Instead, they will indicate some degree of compromised speech. One might legitimately ask if we are missing something, perhaps a subtle notion intended by the Hebrew author related to a wider range of possibilities. Certainly, the idea of circumcision (or uncircumcision) links Moses' lips to other covenantal language, although the *how* and *why* of the relationship are somewhat of a puzzle. To the wicked, God says, "What right have you to recite my statutes, or take my covenant on your lips?" (Ps 50:16). Is it possible that the Hebrew author was trying to tell us that Moses thought of himself as less than worthy to speak out in the Lord's name? In that case, being a poor speaker may have been of secondary importance.

We tend to think of Moses as one of the greatest figures in all of biblical history. That being the case, why would he think of himself as inadequate to the task God assigned for him? If that seems a little difficult to swallow, consider the parallel reaction in Isaiah's response to his call. He says, "Woe is me! For I am lost; for I am a man of unclean lips, and I dwell in the midst of a people of unclean lips; for my eyes have seen the King, the Lord of hosts!" (Is 6:5). Similarly, Jeremiah thought of himself as too young for the task, but the Lord responded, saying, "See I place my words in your mouth!" (Jer 6-9). Thus it is not unusual for a great prophet to think of himself as inadequate for his calling. That includes Moses, whose "uncircumcised lips" are not all that different, it seems, from Isaiah's "unclean lips." Both can be seen as examples of abject humility in the face of an awesome mandate from God. If that be the case, the literal translation preserves more of the Hebrew writer's original intent. Having Moses say simply, "I am not an eloquent speaker," could be construed to be like an excuse to avoid doing what the Lord has asked of him. The expression, "I am of uncircumcised lips," however, may be interpreted as a sign of humility on the part of Moses. This interpretation, at least, would be closer to that of Isaiah's "unclean lips."

There is often a disconnect between the lips and the heart (Is 29:13; Ez 33: 31; Ps 12:2; Prv 20:24; Mk 7:6). The pericope from Mark's gospel is in the context of Jesus' reaction to the Pharisees and scribes, after they criticized the disciples of Jesus for not washing their hands before eating.

> Well did Isaiah prophesy of you hypocrites, as it is written, "This people honors me with their lips, but their heart is far from me; in vain do they worship me, teaching as doctrines the precepts of men." You leave the commandment of God, and hold fast the tradition of men (Mk 7: 6-8)

The quotation in Mark is from Isaiah (29:13). In other instances the disconnect between heart and lips implies not so much hypocritical behavior as deceitful conduct (cf. Ez 33:31; Ps 12:2). Like the mouth, the tongue can be a deadly weapon, especially when used as an organ of deception (Jer 9:3; 9:8). There is even a proverb (25:15) that tells us that a "soft tongue will break a bone."

One of the most remarkable images concerning the tongue, however, is that of the tongue which "cleaves to the roof of the mouth" (Ez 3:26; Jb 29:10; Ps 137:6). In the Ezekiel verse the connotation is that the Lord will make the prophet dumb, or unable to speak. By contrast, the psalmist requests that his tongue cleave to the roof of the mouth if he forgets Jerusalem. The expression, "cleave to the roof of the mouth," means that the tongue is tightly bonded, or stuck to the palate. It is a biblical expression that can be easily tested on oneself. Try saying, for example, with tongue held tightly against the palate (the roof of the mouth), "I am the person who forgot Jerusalem."

While on the subject of garbled speech, there is another kind of incoherent language referred to as "speaking in tongues" (Greek, *glossolalia*). Paul devotes a major portion of chapter fourteen of 1st Corinthians to this subject. He does not deny that it is a gift of the Spirit, but he much prefers someone to prophesy, rather than speak in tongues. When a person speaks in tongues he speaks to God and not to men, "for no one understands him" (14:2). Evidently, Paul is not impressed with one's ability to speak in tongues, for he says, "he who speaks in a tongue edifies himself, but he who prophesies edifies the church" (14:4) Paul even warns, "If, therefore, the whole church assembles and all speak in tongues, and an outsider or unbeliever enters, will they not say that you are mad?" (14:23).

CHAPTER SEVEN

Hand and Arm

> *So the other disciples told him, "We have seen the Lord." But he said to them, "Unless I see in his hands the print of the nails, and place my finger in the mark of the nails, and place my hand in his side, I will not believe."* (John 20:25)

The hand and arm, of the upper extremity are among the most expressive parts of human anatomy. Indeed, there is none other to compare with it excepting the head and face, along with the muscles of facial expression. What would we do without the hands to emphasize a point of conversation? How could we properly love our spouse without the hugs, in the absence of the tender touching of hands and fingers? How unsatisfying it would be if we could not hold our children and grandchildren tightly, and feel their tiny arms about our necks. So important are the arms and hands that we can hardly imagine what it would be to live without them.

Among those with thalidomide birth defects, many born without upper extremities, it is known that some of these children have learned to brush their hair and teeth, holding a comb or a toothbrush between the toes of their feet. Some have become artists, painting a canvas with the same extraordinary effort and determination. It is marvelous how we can adapt to misfortune if we are pressed hard enough. As adults, however, those of us born with all our normal anatomy intact, we would find it almost impossible to cope with the loss of both arms and hands. Whether through disease or accidental causes, such a loss would devastate the ordinary person.

The power of the hand for creativity is awesome. We would never have risen to our elevated state in the animal kingdom in the absence of the hand with its opposable thumb. It is the instrument that gives substance to our creative juices, our intellectual dreams and schemes. It is little wonder that the Bible, among its varied literary forms, gives such importance and prominence to the hand. To deliver a person "into the hands" of another is to put them under the power of another. To "break the arm" of a tribe, or a nation, is to destroy their armies, or render them powerless. Biblical examples of the symbolic use of arm and hand are so numerous that one need not read many pages to find them.

In the Pentateuch, most of the symbolism involving the arm is in reference to the power of the "outstretched arm" of God (Ex 6:6; 15:16; Dt 4:34; 7:19; 9:29; 11:2). In Exodus the Pharaoh would not let the people go unless compelled by "a mighty hand" (3:19). Of the ten plagues the Lord sent into Egypt to show his strength, seven of these involved the power vested in the hand (7:17; 8:6; 8:17; 9:3; 9:22; 10:12; 10:21). During the exodus itself, Moses stretched out his hand over the sea to divide it (14:16), and did the same to cause the waters to return to their normal condition (14:27). In the battle with Amalek, Israel prevailed whenever Moses "held up his hand" (17:11). Thus, the hand and arm, as the symbolic wielder of power, is derived in most cases, directly or indirectly, from the "mighty hand and outstretched arm" of the Lord.

The upper extremity is also a means of demonstrating authority, healing the sick and disabled, or it can be raised to the heavens in supplication. The hand and arm display evidence of just about every human emotion one can imagine. As such, they function as anatomical extensions of the muscles of facial expression. It is the instrument of showing good faith and friendship, or of sealing a contract. In the courts of law we place our hand on the Bible and swear to "tell the truth, the whole truth and nothing but the truth." Used as an instrument of oath taking, the hand becomes the symbol of the most profound and unwavering sincerity one can offer. An example of such a symbolic gesture from the Bible is given below.

Hand Under the Thigh

> And when the time drew near that Israel must die, he called his son Joseph and said to him, "If now I have found favor in your sight, put your hand under my thigh, and promise to deal loyally and truly with me. Do not bury me in Egypt."

(Gn 47:29)

In view of the great dignity and symbolic power associated with the hand, it is not surprising that this anatomical extremity should be the instrument of oath taking. In the lead quotation of this chapter (Gn 47:29), Jacob (called Israel) asks his son Joseph to "put your hand under my thigh." The occasion took place as Jacob was close to death, and the last wish of the dying man was "not to be buried in Egypt." Instead, he asks Joseph to swear that his remains would be taken to the land of his ancestors and buried there.

This tradition is first mentioned in the book of Genesis when Abraham asks his head servant to place his hand under Abraham's thigh. In this case the promise extracted from the servant is to find a wife for his son, Isaac, not among the Canaanites, but among the kindred of his own land (Gn 24:2). The servant was told to swear by the God of heaven and earth that he would do as Abraham demanded. This was how Rebekah came to be chosen as the wife to be of Abraham's son, Isaac.

The question is, of course, what is the modern significance of this ancient tradition? Can it be likened, for example, to placing the hand on the Bible in a court of law? That might seem to be an acceptable comparison, but why "under the thigh?" The practice is perhaps a euphemism for putting the hand in close proximity to the male genitals, the "fountain of reproductivity" (cf. *The New Oxford Annotated Bible*, note on Gn 24:2). Thus the hand, one of the most honored symbols of power, is placed close to the male source of human offspring, before the sacred oath is taken.

Anatomically speaking, a hand placed under the thigh of a man, especially one in a reclining position, would be closest to the testicles (testes). One could say therefore, that Joseph swore on the testicles of his father, Jacob, and it would not be an inaccurate statement. In view of this symbolic act it is not surprising that the Latin word for testate (having left a valid will) means "to bear witness," or to "testify" (*testificare*). Likewise the Latin word for "testament" (*testamentum*) means a proof, a testimony, or to bear witness. All these meanings bear a relationship to the source of human spermatozoa ("seed"), the male contribution to fertilization made in the testes. The conclusion seems inescapable that "placing the hand under the thigh," is not unlike, in the modern context, of placing the hand on the Bible in a court of law.

While on this subject, it should be said that the ancient

Hebrews knew almost nothing about the mechanism of reproduction, beyond the act of sexual intercourse. The men of biblical times believed they provided the "seed" which would bring about a new life in the womb of the mother. They knew nothing of the woman's contribution to the process and were only guessing at their own.

> And Adam knew his wife again, and she bore a son, And she called his name, Seth, for she said, "God has appointed to me another seed in place of Abel, because Cain killed him." (IB, Gn 4:25).

The literal translations are especially needed in this case to understand the word, "seed" in its proper context (cf. Isaiah 53:10). What is certain is that men of biblical times only guessed at their role in procreation, perhaps by analogy with the agricultural process. One plants a kernel of corn, a seed, and a stalk of corn grows out of the earth. With that view in mind it seems unlikely that the patriarchal view of the women's role in reproduction was little more than to nurture the "seed" growing in her body, much as the earth nurtures the growing kernel of corn.

This was not always the case, writes feminist author Elizabeth Fiorenza (*Womanspirit Rising*, 143). "In cultures and periods when the mother was the only known parent," says Fiorenza, "and her pregnancy was easily attributed to the wind or to ancestral spirits. Among primitive peoples, therefore, the power of women to create life must have seemed awesome. Joseph Campbell (*The Masks of God*), suggests that the power of women, viewed as creators of life in those earliest of times, must have been understood as a magical force. It gave women prodigious powers.

In the very earliest of artistic efforts, the so-called "Venus of Willendorf," a woman with swollen breasts and belly, and huge buttocks, may have been one the first examples of a "graven image." Campbell believes these may have been the first objects of worship and religion, writes Fiorenza. Later on, but still in prehistoric time, a knowledge of the role of men in procreation came into being. When this primitive understanding of paternity, "which women must have discovered first," says Morton (*Womanspirit Rising*, 161-162), "the men tried to usurp women's part in the birthing process. They established themselves as the sole parents, reducing women to nurturing 'their seed'."

Given there is some truth to Morton's assessment, it would

probably account for the inferior social positioning of women from that point in history onward. It wasn't until the 18th Century that any notion of the sperm and ovum in procreation were revealed by the Italian researcher Lazaro Spallanzani (1729-99). Prior to that time the mentality of the male as the only source of reproductive seed persisted. In the late 17th Century, for example, not long after the invention of the microscope by Antony van Leeuwenhoek (1632-1723), Nicholas Hartsoeker described a homunculus (little man) in the head of a spermatozoon. Hartsoeker's homunculus contributed to the then common assumption called the "theory of preformation." In other words, the "little man" in the head of the sperm need only to grow and mature in the mother's uterus to become a newborn human being. This may have been the prevailing view among the biblical patriarchs, as many believe, with respect to their "seed." They did not need microsopes to make this assumption. Otherwise the status of the female in those ancient times would have faired much better. Certainly, no one ever swears on the ovaries of the ancient women of the Bible.

The Power and Dignity of the Right Hand

> Then the King will say to those at his right hand, "come, O blessed of my Father, inherit the kingdom prepared for you from the foundation of the world..." (Mat 25:34)

There seems little doubt that the ancient Israelites were predominantly right-handed. Biblically speaking, there is such an emphasis of the right hand, as to its power and importance, one is forced to this conclusion. If the opposite were true, i.e., that most of the ancient people of the Bible were left-handed, it would be hard to imagine that such great dignity would be vested in the right hand. Left-handedness was unusual enough to point it out in one of heroes of the Bible, the Benjaminite, Ehud, who is specifically identified as being left-handed (Jgs 3:15).

It is much better to be blessed by the right hand than the left (Gn 48:18), just as the right hand of God is glorious in power (Ex 15:6; 15:12). In the ordination of Aaron and his sons, the blood of a ram was touched to the right ear, the thumb of the right hand, and the great toe of the right foot (Lv 8:23). A person is cleansed of leprosy in the same manner (Lv 14:14-17), with the use of sacrificial blood and oils.

The Psalms are so filled with references to the special dignity and importance of the right hand that it hardly seems necessary to mention them all here. As two examples, the right hand of the Lord is exalted (118:16), while in another Psalm (137:5), there is a petition for the right hand to wither if Jerusalem is forgotten. The implication here is that it would be a much greater punishment, or loss, if the right hand, rather than the left hand, were to wither. Among the prophets the right hand is no less important. Isaiah tells of how the Lord has "sworn by his right hand and by his mighty arm" (Is 62:8).

In the New Testament the Lord will "place the sheep at his right hand, but the goats at the left...then the king will say to those at his right hand, 'come, O blessed of my Father, inherit the kingdom prepared for you from the foundation of the world...'" (Mt 25:33-34). In the gospel of Mark we are told, "you will see the son of man seated at the right hand of Power (14:62; cf. Lk 22:69; Rom 8:34).

The Hand in Action

There are at least two physical actions of the hand that are symbolically important. The first is the "laying on of the hand(s)," and the other is to "stretch out the hand(s)." In either case the action has significant implications of power and meaning which transcend the physical gesture itself. Placing the "hand under the thigh," also a symbolic act of the hand for purposes of oath taking, was considered earlier.

To Lay or Place the Hand(s).

The book of Leviticus is replete with acts of laying on of the hand or hands upon something, such as on the head of the sacrificial offering (1:4; 3:2; 4:4; 4:24). In the sixteenth chapter, Aaron is instructed to:

> lay both his hands upon the head of the live goat, and confess over him all the iniquities of the people of Israel, and all their transgressions, all their sins; and he shall put them upon the head of the goat, and send him away into the wilderness by the hand of a man who is in readiness.
> (Lv 16:21)

This action becomes for us the origin of the expression, "scapegoat." Thus, anyone who is found guilty of an offense for the sake of allowing others to escape the accusation, is a "scapegoat." The person so labeled may or may not be guilty of the offense of which he is accused. Usually he (or she) proclaims innocence, while invoking the "scapegoat" image as a defense.

In a more ultimate, cosmic sense, Jesus may be viewed as the Christian scapegoat, just as he is identified with the Passover lamb. As the Passover lamb, as well as the scapegoat, Christ is an innocent victim, but in this case the victim freely accepts his destiny, as imaged in the suffering servant oracles of Deutero-Isaiah.

> Yet it was our infirmities that he bore, our sufferings that he endured, While we thought of him as stricken, as one smitten by God and afflicted. But he was pierced for our offenses, crushed for our sins, Upon him was the chastisement that makes us whole, by his stripes we were healed.
> (*NAB,* Is 53:4-5)

In the book of Numbers (Nm 27:18), the Lord instructs Moses to "lay your hand" on Joshua, thus transferring power and leadership to the man who would lead the Israelites across the Jordan and into the Land of Promise. In the early Christian church the tradition of "laying on of hands" became highly significant, although the existence of the practice from the earliest beginnings of the church is questionable. That is to say, there is no clear indication of a separate rite for confirmation (J.N.D. Kelly, *Early Christian Doctrines*, 195).

During the Pauline era, baptism is the vehicle for conveying the Spirit to believers. Paul indicates the importance of baptism as one of sharing in the death of Christ.

> Do you not know that all of us who have been baptized into Christ Jesus were baptized into his death? We were buried therefore with him by baptism into death, so that as Christ was raised from the dead by the glory of the Father, we too might walk in newness of life. For if we have been united with him in a death like his, we shall certainly be united with him in a resurrection like his (Rom 6:3-5).

Later on, there is a tendency to interpret the effect of baptism as one of the "remission of sins and regeneration." Around AD 256, there occurs a further demarcation between baptism and the laying on of hands as "lessor" and "greater" respectively. With the weakening of baptismal significance, therefore, the "imposition of hands or of sealing with chrism," reserved for the bishop, became enhanced. According to Cyprian, confirmation is what bestows the Spirit and apparently the remission of sins as well (Kelly, p. 210).

In the Roman Catholic Church of today, the effect of confirmation is regarded as the "full outpouring of the Holy Spirit as once granted to the apostles on the day of Pentecost." It carries with it an increase and deepening of baptismal grace. Like baptism, which it completes, confirmation is given only once, "for it too imprints on the soul an *indelible spiritual mark*" (*Catechism of the Catholic Church*, 1302-1304).

To Stretch Out the Hand(s).

The outstretched hand is, in most cases, a sign of power. In other cases it is an act of supplication. God repeatedly instructs Moses to stretch out his hand to demonstrate his power to Pharaoh and to make his mark in Egypt (Ex 8:6; 8:17; 9:3; 9:22; 10:12; 10:21). Solomon stretches out his hands to heaven in thanksgiving for the blessings of the Lord on Israel (1 Kgs 8:54). The power of the outstretched hand is again demonstrated as Jesus heals a leper (Mt 8:3; Mk 1:41; Lk 5:13). In another instance of healing, Jesus has the man with a withered hand stretch out his hand so that it can be made whole (Mk 3:5). The withered hand has no power, but the act of outstretching it represents a entreaty to have its normal power restored.

In some cases the outstretched hand seems unrelated to power or healing, nor is prayer and supplication mentioned. When Jesus stretches out his hand toward his disciples, for example, he informs them that those who do the will of his Father are his brother, sister, and mother (Mt 12:49-50). Upon surface examination of these verses there seems to be no indication of power in the gesture. Upon closer study, however, one can see this particular example of the "outstretched hand" as a symbol of authority. In making his pronouncement, Jesus doesn't refer to some ancient authority, such as one of the prophets. Rather, he simply makes the statement, and in so doing, he points to himself as the authoritative figure.

Similarly, when Agrippa gives Paul permission to speak in his own defense, the apostle "stretches out his hand" as he begins to speak (Acts 26:1). In this instance, Paul gives an authoritative account of his conversion experience. He told King Agrippa all that had taken place on the road to Damascus, as no other person could report in the detail or with greater authority than that of the apostle himself.

The outstretched hand, in these cases (Mt 12:49-50; Acts 26:1), therefore, seems to carry the implication of one who speaks with authority. As a corollary, the person of authority also is in a position of power. One does not speak with authority unless there is a sense of power attached. The power of authority does not imply the same level of strength and intensity, perhaps, as does the power to heal illness or to forgive sins. It is power of a sort, nevertheless.

The Kingdom at Hand

An event that is to come in the near future is said to be "at hand." For the day of the calamity "is at hand" (Dt 32:35), or "the kingdom of heaven is at hand" (Mt 10:7; 4:7). Similarly, "its time is close at hand and its days will not be prolonged" (Is 13:22). The question arises, of course, just how soon does "at hand" mean? This resolution of this question seems to be of special concern among biblical scholars.

The kingdom is "at hand" (Mk 1:15; Mt 3:2; 4:17; 10:7) is similar to other usages which imply that the kingdom is close by, right next to us, or at least, not far from us. Similar to "at hand" is the expression that the kingdom has "come upon you" (Mt 12:28; Lk 11:20) or "has come near" (Lk 10:9,10;) or "is near you" (Lk 21:31). All these expressions convey a sense of nearness or contiguity, and to be contiguous with something is to share a boundary with it, or to touch it.

But if the kingdom is "at hand," or is "near," what do these terms have to say about the present and future? Undoubtedly there is a certain tension between the future and the present with respect to the proximity of the kingdom. While the imminent nature of the kingdom is reflected in the "at hand" or in being "near," it is nevertheless seen as belonging to the future, as confirmed by the familiar petition, "Thy kingdom come." By contrast, when Jesus is asked by the Pharisees when the kingdom of God is to come, he responded, saying, "The kingdom of God is not coming with signs to be observed; nor will they say, 'Lo, here it is!' or 'There!', for behold, the kingdom of God is in the midst of you" (Lk 17:21).

Despite the apparent contradiction of the kingdom being "at hand" and at the same time being "in the midst of you," these two referents fit nicely into the heart paradigm (see Chapter Two, also my book on The Biblical Heart). The kingdom is in our "midst" because it is in the hearts of men. That is, the kingdom of God is at hand because it is both *imminent* and *immanent*. It is also"coming" because those of us among the living, at least, have not entered into the full glory of the kingdom. I have maintained all along that the Hebraic sense of the heart must possess something of the divine dimension, because among other reasons, it functions as a conduit between God and his human creatures.

If we add to this notion the idea that the heart represents the temple of the body in which the Holy Spirit resides (Rom 5:5), we have further evidence of a "down payment" on the glory that is to come when we are finally and fully incorporated into the kingdom. As Brevard Childs asserts (*Biblical Theology of the Old and New Testaments*), "the kingdom of God has not come in its glory, but its powers are already at work."

I realize that the conclusions reached here are not the ultimate scholarly word on the subject of the kingdom being at hand, nor do they even come close to the thoroughness required of such an undertaking. My idea in relating the kingdom to the biblical heart, especially as nuanced by the words of the Jesus tradition, is to shed light from an often neglected perspective. Too often the biblical heart is seen as an unimportant metaphor for something else (love, for example). In fact, the heart contributes a spirtual trajectory which moves from Old to New Testaments, providing a continuity between the two traditions not often understood or appreciated fully.

Other Common Expressions of the Hand

There are other interesting expressions found in the Bible, many of which may be the origin of similar ones in modern parlance. To do something in a "high handed" way, is a biblical expression which indicates a certain level of arrogance. One who sins inadvertently is not to be held nearly as accountable the one who sins "with a high hand" (RSV, Nm 15:30). The less literal *NAB* translates this verse as "one who sins defiantly."

To "hand over" something or someone is another expression we have in common with the ancients (Dt 19:12; cf. 2 Kgs 12:7). It means, of course, to release possession of someone or something to another. The Bible also contains some hand gestures that are familiar to us. To "lay my hand on my mouth," as did Job (40:4), is to not speak, that is, to hold the words as prisoners in the mouth. Or as proverbs admonishes, "If you have been foolish, exalting yourself, or you have been devising evil, put your hand on your mouth" (Prv 30:32).

Finally, the hand can serve as an expression of one's compassion, or lack of it.

> If there is among you a poor man, one of your brethren, in any of your towns within your land which the Lord your God gives you, you shall not harden your heart or *shut your*

> *hand* against your poor brother, but you shall *open your hand* to him, and lend him sufficient for his need, whatever it may be. For the poor will never cease out of the land; therefore I command you, You shall *open wide your hand* to your brother, to the needy and to the poor, in the land
> (Dt 15:7-11, my italics).

Thus, the open hand is analogous to an open heart, the soft, or circumcised heart, whereas a closed hand may be compared to a hardened heart, or the heart in need of circumcision (cf. Chapter Two).

CHAPTER EIGHT

The Lower Extremity

> *How beautiful upon the mountains are the feet of him who brings good tidings, who publishes peace, who brings good tidings of good, who publishes salvation who says to Zion, "Your God reigns."* (Is 52:7)

How Beautiful Are the Feet

To be alive and to be meaningful in every age, theology has to consist of a growing appreciation of reality, material and spiritual, human and divine. Reality does not change, as far as we know, but the human ability to conceive it and to be transformed by it, does change. What better symbol of these changes than the lower extremity, the thigh, leg and foot. It was the feet of God's people that took them into Egypt. Their feet brought them out of slavery under the leadership of Moses, then took them on a forty-year trek through the desert before they finally arrived at the Land of Promise. Joshua led the Israelites across the Jordan and into Canaan, where they struggled to gain a *foothold* in the new land. The Israelites were an army on foot, fighting against other foot soldiers. Even horsemen and the drivers of chariots, in the more sophisticated battles, could not dispense with their legs and feet. Had it not been for the ability to move about in those ancient times, there would have been no biblical history. The ability to change locations, physically and symbolically, is an essential ingredient of progress, spiritual and well as physical.

The person who believes that the religious consciousness of Abraham, or even of Moses, was equal to that of an Isaiah, or Jeremiah, or Ezekiel, must see the study of history as a meaningless undertaking. One could argue the position of meaningless history, of course, but the pursuit of this line of reasoning leads, sooner or later, to intellectual bankruptcy. In my opinion there must be purpose for our existence in time, for enduring the sufferings that nature brings as well as those we bring upon ourselves.

We might also speculate concerning the truth of this principle as it may apply to the growth and maturity of the human psyche. That is to say, does psychic development in the human individual in some ways parallel the evolution of human consciousness in history? Could one, for example, observe the mental growth of a child, from infancy to adolescence, from childhood to adulthood, and see a similarity with the mental and spiritual development of our forebears? Did early humankind experience developments in consciousness similar to the growing child? Could we possibly compare the innocence of a young child with the innocence of Adam? I believe that such comparisons are not only possible but may be useful as well. What other reference point, other than the innocence of a child, would we compare with the innocence of Adam?

Sooner or later, of course, we have our personal repertory of temptations and, as a consequence, are "forced out of the garden." Like Adam (or Eve) we want to be our own god, and when confronted with personal disobedience we place the blame anywhere but where it belongs. In this mythic tale, Adam even blames God, as he says (in paraphrase), "It was that woman you put here that made me eat of the fruit" (cf. Gn 3:11). So it is with us. We leave the "garden" of childhood innocence and confront all the problems and temptations which are familiar to us from the numerous stories and examples from the Bible. In the process, the human psyche begins its slow and often painful evolution in religious consciousness.

Saving History Begins.

The lower extremity, particularly the feet, continue here to symbolize motion toward salvation. To consider saving history at this point is appropriate, since the strategy of God always involves motion, whereas sitting still implies stagnation.

Saving history begins in earnest when one of God's creatures develops a sense of "I"ness, or at that time when the living organism could step outside of a singular existence and reflect upon itself. Teilhard de Chardin speaks of this moment of terrestrial evolution, as an ascent of life folding in upon itself, which comes "finally to center itself under our eyes," (*The Phenomenon of Man*). In a similar vein, Jesse

Jackson keeps reminding his audiences, "you are somebody."

Saving history must have begun here, as our hominid ancestors realized for the first time that "they were somebody," and with that realization came the first efforts to account for human origins. We are a superbly curious species, forever wondering where we came from and where we are going. The first time this special faculty of wonderment came into play is the moment that this peculiarly human faculty of the spirit, that which we are calling the psyche (the *heart*, in biblical terms), was created, allowing us to see ourselves in the reflected image and "likeness of God."

Turning to biblical applications of this phenomenon, it is possible, I believe, to see a growing appreciation among the Old Testament writers for humankind's relationship to God. That is, there seems to have been a growing awareness on the part of the biblical authors as to the essential meaning of human creatures and the God who created them; so much so that they had to finally ask, "what are human beings that you are mindful of them?" (cf. Ps 8:4).

The Journey Begins.

At the Passover meal the people are told to eat with their "loins girded, your sandals on your feet, and your staff in your hand" (Ex 12:11). This was so the Israelites would be prepared for a long journey, one that will take them not just through alien lands, but, more importantly, immerse them in a long and arduous spiritual journey as well. From that time onwards, "gird your loins" often can be read as a metaphor for being prepared, to be ready for whatever is to come (cf. Jb 38:3;40:7). Ideally, faithfulness will girdle the loins, as in the case of the Messianic King (Is 11:5). Prophets, especially, must take care to "gird their loins" in preparation for what the Lord will ask of them (Jer 1:17).

During the desert experience, the Israelites, under the guidance of Moses, moved from paganism to monotheism. At this stage in their spiritual development, the people were not unlike rebellious teenagers. And yet the experience of those complaining, self-centered people of God is often thought of as representing a spiritual journey we all experience, a period in our lives of seemingly aimless wandering, a stage of growth in which we are all spiritual teenagers. But at any point in this journey the faithful need only to cry out for help, to express their dependency on God, and they will be delivered from their afflictions, saved from the arid wilderness of spiritual starvation.

The Israelites did cry out to the Lord, and as a consequence, eventually made it to the promised land. Nevertheless, the accomplishment of arriving at the land of "milk and honey," the land of adult perceptions (by analogy), does not mean that the trials are over. It simply means that circumstances have changed. In our personal "land of promise" there will be new challenges which, if accepted and responded to, will help us to grow closer to the goals God has set for each of us in the grand strategy of salvation history. There will undoubtedly be "ups and downs" in the process of spiritual growth as we are presented with new challenges.

The monarchy certainly presented the Israelites with a new challenge, one which almost defeated them, but they grew and matured because of it. In the end, Israel survived a shattered monarchy, and through the efforts of great prophets such as Isaiah, Jeremiah and Ezekiel, grew to higher levels of religious consciousness.

A Stressful World.

Arnold Toynbee (*A Study of History*), makes a comparison with regard to secular history (if there is really such a thing), which I believe applies equally to the history of salvation. He relates progress in civilization to the point on the rim of a wheel. If we follow that point along the course of a vehicle on the way to its destination, we can see that there are "ups and downs" described by the point, but it always moves steadily forward as well. If we were to create the world in our imagination, could we do it any better than God? Could we make the point on the wheel of saving history go straighter without all the "ups and downs?"

This business of playing God and creating an imaginary world is one of my favorite class exercises. It has a certain shock value, to say the least, to ask students in a Scripture class to pretend they are God and to rewrite human history based on their experiences. The problem is this: how to create humanity and avoid all the pitfalls of human frailty we observe in the world today. In the end, the students come to realize that, "as God," they can create robot-like creatures that would always obey God, but such creatures would never be capable of freely returning God's love. Freedom, therefore, is a necessary consequence of our humanity, which by its very nature, opens creation to the possibilities of disobedience and evil.

Neither could we create a world without physical stress, i.e., storms, earthquakes, floods, etc., without circumventing God's plan. Human beings, no matter how unfortunate it may seem, need physical and mental challenges. Toynbee has demonstrated this phenomenon as well. Where civilizations have progressed in certain geographic areas beyond other contemporary populations, such as the remarkable civilization of the ancient Greeks, they did so because of the presence of an optimal amount of climactic and terrestrial stress. Toynbee showed that Human beings without stress, who live in an idyllic environment, a paradise where food is for the picking, and the temperature warm and mostly ideal, make little progress, and are lacking in inventiveness and creativity.

By analogy, the effects of stress and adversity in our physical surroundings applies equally to our spiritual growth. The Bible is ample proof of this. We need challenges to our faith, just as God tested the faith of Abraham, just as the Israelites were challenged in the desert and in the land of Canaan. Without adversity and challenge, it is part and parcel of our human nature to become stagnant, and cease to grow.

As for my students, they usually end up concluding, some reluctantly, that God could not have created the world in any other way. They eventually saw the Lord's wisdom in all things, even in the storms, because without storms in our lives (emotional as well as meteorological ones), we would never pray to be delivered from them. That we are placed in a world of tribulation **because God loves us** is at bottom a difficult idea to accept. It is perhaps one of the most problematic of all aspects of one's faith, but it is through this milieu that the individual comes to recognize his/her need for God, and it is how God operates through saving history to bring us closer, albeit ever so gradually, to a relationship of trust and love. These thoughts and ideas are not original with me, of course, but represent a general consensus among biblical scholars.

The Lessons of the Exile.

With the threat of exile and its eventual fulfillment, the Israelites faced another crisis in their relationship with Yahweh. They had lost the monarchy and they could no longer seek God in their holy Temple in Jerusalem. Even their corporate sense as a nation was fading from memory. It was indeed a time of crisis for the people of God, but one in which history raised up the great prophets, some think the greatest of all prophets, Jeremiah and Ezekiel. It was these two men of God who served as catalysts for a new dimension in the faith of the Israelites. Israel was to be given a "new heart" and a "new spirit" to know God (Ez 18:31). With their old stony hearts in place they could no more change than could "the leopard his spots" (Jer 13:23).

The exilic experience, I believe, has a parallel in the spiritual growth of the individual. As a person becomes more "sophisticated" in his/her religious outlook, progress is always a matter of giving up something to gain something else. What is gained is a more deeply centered response to the call to faith. Christians, therefore, must at times experience an "exile" of sorts, giving up some "sacred temple" of their beliefs, perhaps in the form of a literalist interpretation of Scripture, to reach a more profound appreciation of the kind of God one sees as Lord. Like the point of Toynbee's wheel, the faithful may seem to go backwards and downward for a time, until at last the realization comes that the Spirit is leading all believers to even greater treasures. This was the kind of enlightenment Jeremiah and Ezekiel foresaw. In the absence of the monarchy and in the destruction of Jerusalem's Temple, the people were being called to a higher and more profound relationship with Yahweh. In those terms the exile may be seen as the dark night before the dawn, reflecting the experience of all humankind who seek the face of God, who truly thirst for righteousness.

Beyond Exile and Toward Fulfillment.

Having seen the wisdom of the exile in the broad outlines of saving history, the door of the world is now opened wide for the terminal period of historical gestation. It is fascinating to me that Trito-Isaiah (3rd Isaiah) should use the imagery of a pregnant womb to describe the expectations of his post-exilic world (Is 66:7-9). It is such an appropriate image for this period of history, especially from the point of view of the imminent birth of Christ. Indeed, the metaphorical womb of saving history merges in somewhat dramatic fashion with secular history, as if to prepare the world for the birth pangs of expectancy.

Numerous forces came together at this most opportune point in history—Hellenism, the great Roman peace (or "Pax Romana"), the Jewish Diaspora—these forces representing the birth contractions. Beyond these forces, and lying hidden beneath them is the psychological groundwork of Jeremiah, Ezekiel and the Isaiahs. That is to say, something momentous is going to happen—God is going to intervene in history in some new and unprecedented way—gird your loins, be prepared for it.

Following the unification of Greece by Philip II of Macedonia, his son and successor, Alexander (356-323), defeated the Persians at Issus in 333 B.C. and took control of their territories. Eventually

Alexander established an empire that reached from Macedonia to the Indus River, embracing the entire Fertile Crescent and extending into Egypt. Upon his death, Alexander's empire was divided among his generals, who came to rule over Egypt and Palestine, among other conquered territories.

The Hellenistic (Greek) influence persisted under Roman domination. The Greek language itself became the language of the educated as well as the ordinary citizen. Local tongues would continue to be used, but the language of business, government and education was Greek.

Under the influence of Hellenism and the relative peace of the Roman empire, the dispersion of Jews (the Diaspora), which began during the Babylonian exile, increased its pace until Jews and synagogues could be found from Alexandria in Egypt to the farthest reaches of the Roman empire, including the city of Rome itself. Before his arrest and imprisonment in Rome, the Apostle Paul had thought of Rome as a stopping-off place, perhaps a headquarters, for further missionary work in Spain. There were, therefore, three important historical factors which set the stage for the spread of Christianity:

1. *Hellenism*, which provided a common language and culture throughout the empire. It would become the language of the New Testament writers.
2. *Pax Romana*, which allowed for peaceful travel throughout the vast reaches of the empire, and no doubt encouraged more wide-spread dispersion of the Jews.
3. *The Jewish Diaspora*, which represented *stepping-stones* for missionary activity. The Apostle Paul, for example would first go to the synagogue in each city he visited and begin his preaching there. This gave him a *toe-hold* from which to launch his Christian appeal. In many cases, because of his Christian message, he would eventually be expelled from the synagogue, but by that time he had already made contacts with Gentiles who would offer him shelter and a base of operations, the first *house churches*.

A New Understanding of Life and Death.

As mentioned in Chapter One, salvation in the Old Testament is not looked upon from the point of view of an afterlife. The good life of Israel is thought of in terms of concrete experience in community with Yahweh, who dwells among the people. In other instances the tone of salvation is militant, and sometimes the prosperity described in salvation seems physically extravagant. One cannot totally dismiss the idea of an afterlife in the Old Testament, however. An examination of Psalm 49, for example, reveals the author facing the universality of death, which comes to the righteous as well as the wicked. In this context the psalmist appears to be looking to the possibility of being rescued from the finality of death. "But God will ransom my soul from the power of Sheol, for he will receive me" (Ps 49:15). Psalm 79 has a similar ring to it as the psalmist writes, "Let the groans of the prisoners come before you; according to your great power preserve those doomed to die" (Ps 79:11).

Certainly by New Testament times there is no question as to the nature of the final enemy, which is death. It is this last enemy that Christ will put "under his feet" (1 Cor 15:25). This change in the perceived relationship of the people to God among the Pharisees, and especially as contained in the traditions of Jesus, represents another stage in the rise of religious consciousness. Once again, just as was the case throughout saving history of the Old Testament, the biblical heart has a major role to play in this new relationship. The "new covenant," the "new heart," and "new spirit" overtakes the chosen people and comes to fulfillment in first century Palestine in the life and teachings of Jesus of Nazareth. From that humble beginning the good news of salvation will spread throughout the known world. As Karl Rahner says of this juncture in time: "The whole movement of this history lives only for the moment of arrival at the goal and climax—it lives only for its entry into the event which makes it irreversible—in short, it lives for the one whom we call savior" (G. B. Kelly, Ed., Karl Rahner, 55).

The Feet in Motion

> ...and I will not cause the feet of Israel to wander any more out of the land which I gave to their fathers, if only they will be careful to do according to all that I have commanded them, and according to all the law that my servant Moses commanded them. (2 Kgs 21:8)

The feet do not move the body, of course, without the participation of the thighs and legs. Most of the figurative language regarding the lower extremity, however, has to do with the feet. Thus, the legs and thighs take a lessor role in the language of salvation as compared to the feet. The lead quote of this chapter (Is 52:7) is a good example of this proclivity. Saying, "How beautiful upon the mountain are the feet," is not to say that the legs and thighs of the messenger are any less beautiful. Simply put, the biblical writers recognize that the feet represent the final translation of any bodily movement. It is the feet that touch the ground. It is the feet which make possible any motion toward or away from some goal or object.

Metaphorically speaking, it is the feet that move toward or away from righteousness. It is the feet that stand on solid ground, or slip on treacherous paths. The foot that slips causes the body to stumble, bringing the person closer to evil or selfish ways. To be under the feet of another is to be under his power. Sometimes it is God's "feet" that wield power over humankind's enemies, and these enemies can be real or abstract.

Examples of movements of the feet may clarify the above discussion. Job (31:5) speaks of the foot which "hastened to deceit." Proverbs has many examples of the foot as metaphor (1:15; 4:27), as does the book of Psalms (119:59). In verse 105 of this psalm the word of God is a "lamp to my feet, and a light to my path." Isaiah (58:13) warns of turning the foot towards selfish ways, and speaks of feet that "run to evil" (59:7). As an example from the New Testament, the Canticle of Zechariah praises God for having been a "guide to our feet into the way of peace" (Lk 1:79).

The Slipped Foot.

A number of references to the "slipped foot" in the Psalms are clearly spiritual metaphors (17:5; 18:36; 94:18), although one could argue that the corresponding physical event cannot be ruled out entirely. A similar case may be made for Jb 12:5 and Prv 3:23. It is more difficult, perhaps, to conclude that Jeremiah is making reference to a physical event when he exhorts, "Give glory to the Lord your God before he brings darkness, before your feet stumble on the twilight mountains, and while you look for light he turns it into gloom and makes it deep darkness" (13:16). It seems highly unlikely that someone would give glory to God solely to prevent a physical slip of the foot. A slipped foot or stumbling feet, therefore, are serious business, biblically speaking. It means one has wandered from the path of righteousness onto the slippery path of sinfulness. The psalmist often gives praise to God for preventing such a fundamental tragedy, for smoothing his path, for keeping his feet on level ground.

Under the Foot.

To have one's enemies under foot is the equivalent of having power over them (1 Kgs 5:3; Ps 47:3; Mt 22:44). Sometimes it is not an enemy as such, but simply the created works of God over which humankind is given dominion (Ps 8:6). A reference to "the enemy being under the feet," however, does not always infer an existing nation, an army, or an individual. It can be an illness, the temptation to do evil, etc., or it could be the "last enemy," which is death. "For he must reign until he has put all his enemies under his feet. The last enemy to be destroyed is death, for 'he subjected everything under his feet'" (*NAB*, 1 Cor 15:25-27).

Washing of the Feet

> So, during supper, fully aware that the father had put everything into his power and that he had come from God and was returning to God, he rose from supper and took off his outer garments. He took a towel and tied it around his waist. Then he poured water into a basin and began to wash the disciples' feet and dry them with the towel around his waist (Jn 13:2-5).

There is an ironic twist regarding the subject of the feet. Often they are seen to possess great honor, even beauty, when they bring good tidings (Is 52:7). When they bear the ark of the Lord across the Jordan, and as soon as the soles touch the water, "the Lord of all the earth shall rest in the waters..." and the "Jordan will be stopped from flowing..." (Jos 3:13). Then, when the feet of the priests touch dry ground on the other side, the waters of the Jordan returned to their place (Jos 4:18).

As a sign of hospitality and genuine welcome, Abraham has water brought to his visitors so that they might wash their feet (Gn 18:4; 19:2). Before one enters the meeting tent or approaches the altar, one must wash his feet (Ex 30:21). Yet, for one to wash another's feet might be thought of as terribly demeaning. Thus, to show her humility, Abigail, acknowledging the great honor bestowed on her by king David, said, "Behold, your handmaid is a servant to wash the feet of the servants of my Lord" (1 Sam 25:41).

Similarly, as a sign of her great humility, the sinful woman washed the feet of Jesus with her tears (Lk 7:38). Jesus himself shocked the disciples by washing their feet at the Last Supper (Jn 13:2-5). In doing so Jesus gave the disciples a model to follow (Jn 13:15), an example of service to one another they should make every effort to emulate. This scene in the gospel of John is a practical extension of what Jesus says in the gospel of Matthew, "The greatest among you must be your servant. Whoever exalts himself will be humbled; but whoever humbles himself will be exalted (Mt 23: 11-12). This gospel account is not unlike the paradoxical statement of Paul (*NAB*, 2 Cor 12:10), "Therefore, I am content with weakness, insults, hardships, persecutions and constraints, for the sake of Christ; *for when I am weak, then I am strong*" (my italics). In both cases (Paul's and John's account), the paradox is that one becomes closer to God, not by promoting one's own importance, but by doing just the opposite. To be important in the kingdom of God, one must humble himself, or to be strong in one's faith, one must recognize his own weakness and dependency on God.

CHAPTER NINE

Flesh and Blood, Bone and Sinew

You clothed me with skin and
flesh,
and knit me together with bones
and sinews.
You granted me life and
steadfast love,
and your care has preserved my
spirit.

(*NRSV*, Jb 10:11-12)

One way for an author to turn a reader off entirely, or, at the very least, risk boring him to death, is to split hairs over the meaning of words. It is with some trepidation, therefore, that I begin this chapter with just such a discussion. I will make it as brief as possible. I justify this risky beginning because I believe it is essential to point out that the Old Testament has no equivalent word for "body." The closest Hebrew word is *basar*, which means "flesh." One can, of course, translate basar as "body," but it could be misleading to do so. Robinson (*The Body*, 11), points out "the remarkable fact" that there is no Old Testament background for soma, the Greek word in the New Testament meaning "body." There is a separate Greek word for "flesh" (*sarx*), the rough equivalent of the Hebrew word, *basar*. How is it that the Israelites, asks Robindon, "made do with one word (*basar*), where the Greeks required two?" The answer to this question, believes Robinson, brings us to some of the most fundamental assumptions of Hebraic thinking about humankind. I will attempt to get to the bottom of these "assumptions" by examining the context of the word, "flesh," in various Old Testament narratives, poetry, and prophetic oracles. From this point on I shall avoid as much as possible the use of Hebrew and Greek words.

The Flesh

> And it shall come to pass afterward, that I will pour out my spirit on all flesh; your sons and your daughters shall prophesy, your old men shall dream dreams, and your young men shall see visions (Joel 2:28; cf. Acts 2:17).

Old Testament Perspective.

The ancient Israelites were not too knowledgeable concerning human anatomy, but they were not totally ignorant of it either. They knew of organs and tissues, bone, blood and skin, for example, from the animals they sacrificed as offerings. Although they did not dissect human bodies, they must have had some acquaintance with human anatomy from the observation of battle wounds. Comparing what little they knew of human anatomy with that learned from sacrificial animals, they were probably able to infer certain conclusions about human structure. It was easy, for example, for the author of Job to distinguish between flesh and the skin which covered it (Jb 10:11). Flesh, of course, now and in the biblical sense, is muscle. Muscle, or flesh, is further distinguished from skin (at least in some cases), and sinew, or tendonous connective tissue. Organs such as kidneys, liver, heart, lungs, etc. were certainly known as vital structures in animals, and by inference, in humans as well. Bones were also common supportive structures in animals and humans alike.

Ordinary experiences would lead them to understand that blood is directly related to life, insofar as loss of a large quantity of it leads to death, whether on the battle field or in the case of a sacrificial animal. Once a broken bone was experienced the community would understand that the bony skeleton was the supporting framework of the flesh, human or animal.

Beyond these simple observations of human anatomy, the Israelites were very skillful at including what they knew of their human composition into their poetry, wisdom traditions, and prophetic utterances. Metaphors abound which relate their emotions and spiritual aspirations to flesh, blood, bone, and various internal organs. In this context, figurative language becomes for the biblical writers an effective device for showing a depth of feeling not otherwise possible.

Wolff (*Old Testament Anthropology*), reports that the Hebrew word for flesh occurs 273 times, and 104 of these instances, or more than a third, are used in connection with animals. This alone, says Wolff, shows that "flesh" is a broad term "characteristic of both man and beast." It is possible to talk of human flesh as food, even, just as with animal flesh. One of the consequences of a disobedient Israel is "you shall eat the flesh of your sons, and you shall eat the flesh of your daughters" (Lv 26:29).

In spite of this commonality, human flesh has a certain dignity that lower animals do not possess. To begin with, man and woman are created in the image of God (Gn 1:26-27), and are given dominion over the animals (Gn 1: 28-30). God makes a covenant with human flesh in the circumcision of Abraham and the males of his household (Gn 17:11-25). In the book of Joel we are told by the prophet that God will pour out his spirit on all flesh (Jl 2:28).

The biblical authors seem to be striving for a balance in their assessment of the flesh. On the one hand, God has placed a certain value and dignity on human flesh, while at the same time, the people are reminded from the perspective of God, that "All flesh is grass, and its beauty is like the flower of the field" (Is 40:6). The same oracle goes on to say that "the grass withers, the flower fades," but in the end, "God's word will stand forever" (Is 40:8).

One of the most positive meanings attributed to flesh is found in the prophet Ezekiel. From the mouth of the prophet speaking for God, we are told, "A new heart I will give you, and a new spirit I will put within you; and I will take out of your flesh the heart of stone and give you a heart of flesh (Ez 36:26; cf. 11:19). Wolff asserts that this positive use of flesh as it relates to human behavior, "is absolutely unique" (p. 29). Chapter Two on the *Hebraic Heart* puts this verse in the context of saving history. It is perhaps difficult, if not impossible, to understand Ezekiel's prophecy in the absence of this context.

To begin with, a heart of flesh is certainly better than a stony one. A hardened heart (or heart of stone) is closed off to God. The person thus afflicted is essentially dead, because he lives only for himself, in selfish pursuit of pleasure. He oppresses the poor and needy, defiles his neighbor's wife, and lifts up his eyes to idols. Such a person, Ezekiel tells us, "shall surely die" (Ez 18:13). A person with a stony heart is not fully human, certainly not authentically so, for he has forsaken the spirit of God and lives only for himself. Compared to this condition, a heart of flesh, one that is not closed off to God's word, is certainly more desirable. "Flesh," in this context, takes on a very positive light.

Authors of the New Testament certainly understood Ezekiel in this light. Here, God's power manifests itself in human weakness. The more one recognizes the limitations of the flesh, one's own proclivities for sin, the more one sees his/her dependency on God. In this sense our weakness in the "flesh" is a positive thing, because it can help us to better see our relationship of dependency on God. That is Paul's meaning when he speaks of his own "thorn in the flesh."

> Three times I appealed to the Lord about this, that it would leave me, but he said to me, "my grace is sufficient for you, for power is made perfect in weakness." So I will boast all the more gladly of my weaknesses, so that the power of Christ may dwell in me. Therefore I am content with weaknesses, insults, hardships, persecutions, and calamities for the sake of Christ, *for whenever I am weak, then I am strong* (my italics). (*NRSV*, 2 Cor 12:8-10)

The flesh, insofar as its human valuation, can be seen as a mixed blessing. In any event, we are stuck with the flesh of the body in this life. It characterizes human life in its weakness and frailty. Mortal humankind is tied to the flesh in contrast to God, who is spirit, according to Judeo-Christian belief. Job is bold enough to ask God, "do you have eyes of flesh?" (Jb 10:4). It is a most intriguing question, even if considered as a rhetorical one. Job seems to asking, "what kind of God are you?" More appropriately, "can a God of spirit care what happens to human flesh?"

Out of this tension between the flesh we humans are made of and the God of spirit we worship grows a desire to be like God. The faithful are inspired to be like God, filled with loving kindness, but most of all we humans thirst for the immortality of God. There are hints of this appetite for immortality in the Old Testament (see Chapter One), but it is never expressed clearly until the book of Daniel (12:2) was written, sometime during the persecution of Antiochus Epiphanes (167-164 B.C.).

The Christian Perspective.

In Christian belief the thirst for a godlike existence of life after death is fulfilled in Christ-event. The hope of immortality is most clearly expressed by the apostle Paul (1 Cor 15:1ff), who asserts that the resurrection of the faithful is assured, because Christ was raised from the dead. He goes on to conclude, "If there is no resurrection of the dead, then Christ has not been raised" (1 Cor 15:13). The Christian perspective on the flesh (*sarx*) is that it is part and parcel of the body (*soma*). The body, in turn, is much more than the sum of its flesh, blood, bone, vital organs, etc., because it is the temple of the immanent God, the indwelling Spirit of God (1 Cor 3:16). Not only that, we are being transformed into something that transcends mere flesh and blood.

> And all of us, with unveiled faces, seeing the glory of the Lord as though reflected in a mirror, are being transformed into the same image from one degree of glory to another; for this comes from the Lord, the Spirit (*NRSV*, 2 Cor 3:18)

As a consequence of this transformation, we have become a "new creation" (2 Cor 5:17; Gal 6:15). To add another dimension to this awesome concept, we are told by the author of 1 John, "Beloved, we are God's children now; what we will be has not yet been revealed. What we do know is this: when he is revealed, we will be like him, for we will see him as he is" (*NRSV*, 1 Jn 3:2-3).

It is texts such as these that led the Greek fathers, such as Irenaeus, to conclude that, "He [God] became what we are in order to enable us to become what He is" (J.N.D. Kelly, *Early Christian Doctrines*, 172). The dominant thread of the thinking of Athanasius, another important church father, is the physical theory that "Christ, by becoming human, restored the divine image in us." Or, to paraphrase the words of Athanasius, "God became human so that humans could become like God," (cf. Kelly, p. 378-379).

The idea of the weaknesses of human flesh can be looked at from a totally new perspective in the New Testament. Human flesh can be thought of as a curse, because it can and does lead one into a life of sin, and sometimes brings on the most contemptible of evil acts imaginable. By the same token, the admission of this weakness of the flesh can bring us closer to saintliness, since it forces us to recognize our dependence on God. Arriving at the conclusion that we are our own worst enemy is the first step to personal salvation. That is why Paul said, as quoted earlier, "whenever I am weak, then I am strong" (2 Cor 12:10).

Flesh and Blood

To the ancient Israelites the blood had no relationship to the heart, at least not in the modern sense. They knew nothing of the heart as a pump for the blood or of its circulation. The blood, however, was the life-giving force of the flesh, human or animal. In this view of the blood as the life of the flesh, it is a sacred substance for use in ordination rituals, and sacrificial offerings.

One must not eat flesh with its blood. To eat flesh with its blood is a sin against Yahweh. Blood must not be poured out on a rocky surface, but upon the earth so that it can be covered with dust. Some of the blood of the sacrificial animal is to be sprinkled by the priest on the altar of Yahweh (Lv 17:6). This ritual expresses the "cultic-theological stringency," says Wolff, that man is entitled to the flesh that comes from the earth, "while the life [the blood] belongs to Yahweh alone" (*Anthropology of the Old Testament*, p. 61). The blood of a murdered man cries out for revenge. The animistic power goes on working even after an innocent man's death, since it finds a hearer in Yahweh (Wolff, p. 61). Thus, the murderer is a fugitive for life.

The Hebrew word for blood, reports Wolff, occurs 360 times in the Old Testament, most frequently in the book of Leviticus (88 times), as one might expect from its constant use in religious rituals. The frequency alone of this word speaks for its importance among the ancient people of the Bible. The association of blood (along with breath) expresses an ultimate concern for life. This reverence for life, asserts Wolff (p. 62), does not stem from the manifestation of life itself, but from the belief that the breath and blood belong to Yahweh.

The Christian Eucharist

The Old Testament traditions regarding flesh and blood make for an interesting, if not provocative, background for the Christian Eucharist. Indeed, the proscriptions against eating blood seem in direct opposition to the central thrust of the Eucharistic sacrifice.

> Only, you shall not eat flesh with its life, that is, its blood. For your own lifeblood I will surely require a reckoning: from every animal I will require it and from human beings, each one for the blood of another, I will require a reckoning for human life (*NRSV*, Gn 9:4-5).

Thus in the old economy, eating the blood of any animal is tantamount to murder, certainly a most grievous sin against Yahweh. In contrast to this proscription against eating blood, John's gospel has Jesus saying, without qualification, "Truly, truly, I say to you , unless you eat the flesh of the Son of man and drink his blood, you have no life in you" (Jn 6:53; cf. 6:54-56). It is no wonder that many of the disciples are reported to say, "This teaching is difficult" (Jn 6:60). Then Jesus, sensing the problem in the minds of the disciples, says:

> Does this offend you? Then what if you were to see the Son of Man ascending to where he was before? It is the spirit that gives life; the flesh is useless. The words I have spoken to you are spirit and life. (Jn 6:61-64)

What Jesus says here is crucial to understanding the central thrust of the fourth gospel. This unadorned statement about eating the flesh of Jesus and drinking his blood is a part of this effort, i.e., to show unequivocally that Jesus is the divine Son of God. The question remains, "how does this blunt statement about flesh and blood relate to the divinity of Jesus?" I'll return to this question momentarily, but first I want to compare Eucharistic imagery in John with the synoptic gospels, Mark, Matthew and Luke.

In the synoptic gospels, at the Passover supper, the circumstances and the mood of the disciples couldn't be more different. In this instance Jesus makes it clear that his "body" and his "blood" are to be taken under the species of bread and wine.

> While they were eating, Jesus took a loaf of bread, and after blessing it he broke it, gave it to the disciples, and said, "Take, eat; this is my body." Then he took a cup, and after giving thanks he gave it to them, saying, "Drink from it, all of you; for this is my blood of the covenant, which is poured out for many for the forgiveness of sins.
> (*NRSV*, Mt 26:26-28).

With minor differences, this is how the last supper is described in Mark, Matthew, and Luke. It is much the same in Paul's first letter to the Corinthians. It is made clear that Christ's body and blood is to be taken in the form of bread and wine, not a piece of flesh severed from the living Jesus, or a cup of blood drained from the veins of Christ on earth.

Finally I believe we can come to some conclusion regarding the unadorned Eucharistic reference in the gospel of John. From the very beginning of the fourth gospel, in the prologue, there is no question concerning the origin of the Son or his divine nature. It is also clear that the Son is pre-existent, for "In the beginning was the Word, and the Word was with God, and the Word was God" (Jn 1:1). "And the Word became flesh and lived among us" (Jn 1:14). This very determined effort to show Jesus' divinity crops up time and again in the gospel itself. One fairly obvious reference to be made is the "I am" (*ego eimi*) statements of Jesus. The context of the "I am" in John's gospel unmistakably refers to Jesus' divinity. As an example, "Truly, truly, I tell you, before Abraham was, I am" (Jn 8:58). Another statement says the same thing in a different context, when Jesus announces to Martha, "I am the resurrection and the life; whoever believes, even if he dies, will live" (Jn 11:25).

Because John's gospel places great emphasis on the divinity of the Son, it is referred to by many scholars as "high christology." That is not to say that the synoptic gospels (often referred to as "low christology"), deny the divinity of Christ. Rather, the disciples are not fully cognizant of this until the post-resurrection appearances.

We return now to the seeming conflict with the Old Testament proscriptions against taking flesh with its blood, and what seems to be John's Eucharistic formula (cf. Raymond Brown, *An Introduction to the New Testament*, p. 346). Recall that, immediately after the disciples complain about his "difficult teaching," Jesus goes on to ask, "Then what if you were to see the Son of Man ascending to where he was before?" Instead of explaining his words about eating his flesh and drinking his blood, Jesus points to his divine origin. In other words, "okay, suppose what I ask of you does seem blasphemous. What if I am from God? Is it all right for the Son of God to ask this of you?"

Now in the Old Testament blood was life, and as such, belongs to Yahweh. The people have dominion over most created things, but not life itself, which God alone controls. Thus, John uses the Eucharistic formula to again demonstrate the divinity of Jesus. The blood that is life belongs to God alone. Therefore, only God has the power to offer his life blood as sacrifice in the Eucharist, and since Jesus is the "Word made flesh," he is fully qualified to freely offer up his life. In this case, because it is the blood of God's divine Son, the life of the blood is eternal.

In my opinion John's gospel makes it clear from the outset that he is relating to the Old Testament tradition. He is certainly not avoiding it when he uses the Greek word for "flesh" (*sarx*), which directly relates to the Hebrew word, *basar*, also meaning flesh. In the synoptic gospels, and in the Pauline corpus, the Eucharistic formula makes use the Greek word, soma, meaning body, a word not found in the Old Testament.

The forth gospel is consistent in demonstrating Jesus's divinity wherever possible. Misunderstanding is a common literary device which often precedes these instances. The narrative account of Nicodemus, for example, tells of a Jewish ruler who misunderstands Jesus' reference to being "born again" (Jn 31-15). Jesus explains what he means by the phrase "born again," (Gr., *anwqen*), then goes on to speak of the Son of Man who has come down from heaven. Similarly, the Samaritan woman does not understand what Jesus means by "living water" (Jn 4:10-11). This incident is used to explain that the living water Jesus gives "will become [in those who receive it] a spring of water gushing up to eternal life" (*NRSV*, Jn 4:14). Then, while still speaking with the woman at the well, Jesus proclaims himself the Messiah (4:25-26).

Jesus' particular invitation to eat his flesh and drink his blood involves a misunderstanding not unlike the stories of Nicodemus and the Samaritan woman. The disciples' reaction, "The teaching is difficult; who can accept it" (*NRSV*, Jn 6:60), prompts Jesus to explain that he is the divine Son come down from heaven, and therefore has the authority to offer his own flesh and blood (Jn 6:62).

Bone and Sinew

> Then he said to me, "Prophesy to these bones, and say to them: O dry bones, hear the word of the Lord. Thus says the Lord God to these bones: I will cause breath to enter you, and you shall live. I will lay sinews on you, and will cause flesh to come upon you, and cover you with skin, and put breath in you, and you shall live; and you shall know that I am Lord." (Ez 37:4-6)

In some ways bones symbolize the opposite of blood. Blood represents life to the ancient Israelites, whereas bones are often associated with death. Bones are what is left after the skin, flesh, and all other soft body parts have decayed and fallen from the skeleton. Sinew is a dense fibrous connective tissue which stabilizes joints (ligaments) and attaches muscle to bone (tendons). These are the last of the non-bony tissues to decay.

The above quote from Ezekiel (37:4-6), is an example of resurrection imagery. Scholars tell us the vision has nothing to do with an actual promise of resurrection. Rather, it is a symbolic reference to God's promise to restore Israel from its condition in exile. Many of the exiles seem to have given up hope of rescue and a return to Jerusalem. They counted themselves among the dead as a nation of Israel.

Bones of the dead were worthy of respect, however, and were often given a special burial site, or moved from one place to another, such as the bones of Israel (Gn 50:13) and Joseph (Jos 24:32), which were brought out of Egypt to be buried with their ancestors in Canaan. In a like manner the prophet asked his sons to lay his bones beside the bones of the man of God (1 Kgs 13:31). Also, miracles are associated with the bones of the dead, as were those of the prophet Elisha, which brought about the resuscitation of a man thrown into the prophet's grave (2 Kgs 13:20). By contrast, The Lord warns that there will be no respect shown to the bones of sinners, those that worship false gods. They will be taken from the grave and laid out on the ground to dry in the sun (Jer 8:1-2).

In the living, bones can be the source of deep emotions, such as fear (Jb 4:14; cf. Jer 23:9). Living bones can also be troubled (Ps 6:2), or they can rejoice (Ps 51:8). I suspect that any deeply felt emotion can be attributed to the bones. Such usage would be analogous in modern parlance when we say, "I feel it in my bones." These expressions may be thought of as an offhand way of recognizing a thought or impulse that has no basis as an intellectual conclusion. It is an intuitive rather than cerebral thought, something "felt" rather than known with certainty. In a similar fashion, the bones are "refreshed" by good news (Prv 15:30), whereas a "downcast spirit" dries up the bones (Prv 17:22).

In light of the above imagery, it is understandable that "a soft tongue can break bones" (Prv 25:15). When the atmosphere is charged with angry emotions, a kind word, softly spoken, can diffuse the situation. Good advice, it seems, put into an effective and graphic image.

CHAPTER TEN

Inward Parts, Outward Beauty

> *For the king of Babylon stands at the parting of the way, at the head of the two ways, to use divination; he shakes the arrows, he consults the teraphim, he looks at the liver.*
> (Ez 21:21)

Liver, Kidney and Bowels

Wolff (*Anthropology of the Old Testament*, p. 64) informs us that in the Akkadian language that the liver is the most important organ after the heart, and it is named in Babylonian literature with great frequency. The liver and "entrails" were often used in divination, practices that were an abomination to Yahweh (Dt 18: 10-12). This would explain why the liver is mentioned so few times in the Bible. It occurs fourteen times in the Old Testament, thirteen of which refer to the livers of sacrificial animals. That is not to say that magic, soothsaying, and divination were not practiced among the Israelites. It would be difficult to imagine, surrounded as they were by these pagan practices, that the People were not influenced by them. If the converse were true, it would be difficult to explain why it is condemned so many times in the Bible (Dt 18:10-12; Is 8:19; 44:24-25; 47:12-15; Acts 8:9-24). These are a few examples, but one in particular I include here.

> I am the Lord, who made all things, who alone stretched out the heavens, who by myself spread out the earth; who frustrates the omens of liars, and makes fools of diviners; who turns back the wise, and makes their knowledge foolish...
> (Is 44:24-25)

Such remarks would have been unheard of in Babylon. There the priests were the judges, lawyers and physicians. This was a logical consequence of the law and medicine, along with theology, all of which were of divine origin. To the Babylonians the liver was the organ that contained more life, or blood, which was much the same thing. This relationship of blood and life, therefore, was shared with the Israelites. In divination all the peculiarities of the liver were noted, its shape, markings, and the form of the gall bladder. Before any business or military venture was undertaken, the liver of a sacrificial sheep would be examined. A clay model of the liver was used as a reference point, which was perforated with holes into which pegs could be inserted. Thus, any anomaly or unusual marking on the fresh liver could be recorded on the clay model by inserting a peg in the appropriate hole (cf. Ralph Major, *A History of Medicine*, p. 32).

One can see why the Hebrew prophets would be against such practices, seeing them as attempts to lay hold of powers which properly belong with the Lord. Although a gradual process, the religion of Israel elevated the people above base superstitions, including the practice of magic and soothsaying. This phenomenon was largely due to the monotheism which came to dominate the thinking of Israel's greatest minds. There were always some who clung to the old practices from pagan times, however, even incorporating them into the religion of Israel. In the Land of Promise (Canaan), as we have seen, fertility cult rituals and the like were commonly practiced by the Israelites, much to the chagrin and frustration of the prophet Hosea.

The Kidneys and Inner Feelings.

I should make a note that the same Hebrew word (*kilyah*) applies to the kidneys (reins) of sacrificial animals and human beings alike. And yet, in the case of human kidneys, the word is seldom translated as such. The *NRSV* as well as the *RSV*, usually very literal in its translations, does not render human kidney in the literal sense. In view of this, one misses the point that Wolff makes (p. 65), that "the kidneys are the most important internal organ in the Old Testament." According to the most literal sense from which Wolff draws, the kidneys are referred to in particular as being created by God (Ps 139:13; cf, *NRSV*, "inward parts"). Human kidneys can also become the seat of conscience (Ps 16:7; cf. *NRSV*, "heart"). Similarly in Jeremiah 12:2, the Hebrew word for kidneys is translated as "heart" in *NRSV*).

There are other examples of this literal sense of *kilyah* that

Wolff brings to our attention. I do not, however, think that Wolff makes a case for the kidneys that differs that much from bones. Deeply felt emotions can also be expressed in terms of the bones. In most instances, I believe, the case could be made that the bones and kidneys alike, being unseen (as inward parts), become metaphors for emotions usually ascribed to the Hebraic heart. Unlike the kidneys, the bones persist after death for an indefinite length of time and, as such, may continue to cry out for respect and honor.

The Bowels.

There seems to be nothing of great theological interest in relation to the bowels or intestines (entrails). These inward parts are mentioned many times in connection with sacrificial animals. In the case of humans, they are mostly named in connection with a gruesome description of a person's death. Two examples will serve: "Joab struck him...and shed his bowels to the ground...and he died" (2 Sm 20:10) and, "...falling headlong he burst open in the middle and all his bowels gushed out" (Acts 1:18).

There is one example in which there is an indirect reference to the lower digestive tract. This example, I believe, has theological significance. Jesus instructs the crowd:

> Then he called the crowd to him and said to them., "Listen and understand: it is not what goes into the mouth that defiles a person, but it is what comes out the mouth that defiles." (Mt 15:10-11, *NRSV*)

Upon further questioning from Peter, Jesus says,

> "Do you not see that whatever goes into the mouth enters the stomach, and goes out into the sewer. But what comes out of the mouth proceeds from the heart, and this is what defiles. For out of the heart come evil intentions, murder, adultery, fornication, theft, false witness, slander."
> (Mt 15:17-18, *NRSV*)

Although Jesus does not directly mention the bowels, he certainly implicates their normal function as a passageway to the sewer. These words represent an attack on the strict legalism of the Pharisees, who neglect the spirit of the Law in favor of their human precepts.

Beautiful Flesh in Saving History

There is nothing beautiful about our inward parts. Their bloody, glistening surfaces are more likely to evince nausea than poetic verse. The opposite is true of the flesh and its covering of skin. The flesh, hung as it is upon the bones give a woman a comely figure, or a man his rugged physique. Cover the shape of the flesh with unblemished skin and you have a beautiful woman, or a handsome man.

Beauty can be a mixed blessing, as it certainly was in the case of Abram (Abraham) and his wife Sarai (Sarah). She was so beautiful that Abram was afraid the Egyptians would kill him in order to possess her. So he asked Sarai to pretend to be his sister. This solved the problem as far as Abram's life was concerned, but it meant that Pharaoh could take Sarai into his house unencumbered by the knowledge she was Abram's wife. The story goes on to tell of Pharaoh's affliction, which presumable occurred before Pharaoh could take sexual advantage of Sarai. One must feel a great deal of sympathy for Sarai, however, to be placed in such a compromising situation because of her beauty (cf. Gn 12:10-20).

Rachel was another beauty of the patriarchal period. It could be said that her beauty played a part in saving history, since she was so lovely in the eyes of Jacob. He eventually married Rachel in spite of the father (Laban) trying to pass off his older daughter, Leah, in place of Rachel. He persisted in his desire for Rachel because his love for her was great. Jacob served the father another seven years because he (Jacob) could not accept Leah as the replacement for Rachel. Despite difficulties in child bearing, Rachel finally gave birth to Joseph who, in turn, figures prominently in saving history (cf. Gn 29:9-28).

Rachel's beauty was passed on to her son Joseph, who was "handsome and good looking" (Gn 39:7, *NRSV*), but Joseph pays for his good looks in much suffering. First his brothers were jealous of him because of his being the favorite of their father, Jacob (Israel). Consequently, the brothers trapped him and sold him into slavery in Egypt. There his handsome appearance attracted the wife of his master, whereupon she said to Joseph, "lie with me" (Gn 39:7). When Joseph refused, the spurned wife accused him of attempted rape. Joseph was sent to prison as a consequence. While in prison he became known for his ability to interpret dreams, and in this way his talents came to the attention of Pharaoh. Joseph's life in Egypt, as a powerful leader under Pharaoh, plays an important role in the saving history of the people. Joseph's beauty, therefore is all the more meaningful because it plays an important role in God's plan for the people of Israel. Eventually there arose a Pharaoh "who did not know Joseph" (Ex 1:8), who enslaved the Israelites, an event which eventually brought about the exodus of the people from Egypt under the leadership of Moses.

David had beautiful eyes and was handsome (1 Sm 16:12). His good looks were one of the reasons he was chosen to serve king Saul (1 Sm 16:18), thus beginning his long, and often tragic relationship with the king of Israel. This association is recorded in the Bible as significant in the history of Israel's monarchy. David, of course, becomes Israel's greatest king and one of the most important figures in the history of salvation.

The narrative about David and Bathsheba is, to me, a good example of how God's saving strategy works through human weakness. David's sin was indeed great, first as an act of adultery, and to make matters even more grievous, the murder of Bathsheba's husband, Uriah the Hittite. It all begins when David, walking on the roof of the king's house, saw a woman bathing who "*was very beautiful*" (2 Sm 11:2, my italics). David could not overcome his lust or his sinful intentions. One thing led to another, beginning with Bathsheba's pregnancy, and ending with the orchestrated death of Uriah.

Sinfulness aside, David was a favorite of the Lord. The parable of Nathan concerning David's lapse is, I believe, one of the most powerful in the entire Bible.

> But the thing that David had done displeased the Lord, and the Lord sent Nathan to David. He came to him, and said to him, "There were two men in a certain city, the one rich and the other poor. The rich man had very many flocks and herds; but the poor man had nothing but one little ewe lamb, which he had bought. He brought it up, and it grew up with him and with his children; it used to eat from his meager fare, and drink from his cup, and lie in his bosom, and it was like a daughter to him. Now there came a traveler to the rich man, and he was loath to take one of his own flock or herd to prepare for the wayfarer who had come to him, but he took the poor man's lamb, and prepared that for the guest who had come to him." Then David's anger was greatly kindled against the man. He said to Nathan, "As the Lord lives, the man who has done this deserves to die..." Nathan said to David, "You are that man!" (2 Sm 12:1-7, *NRSV*).

Absalom's beautiful sister, Tamar, is another tragic story of lust, betrayal, and revenge. It all begins with Amnon's desire to lie with his sister. After the deed was carried out, Tamar was shamed beyond measure. Absalom then plotted the murder of his brother Amnon, which his servant carried out following his instructions. This tragic story made such an impression on William Faulkner, that he was inspired to write one of his most famous novels, *Absalom, Absalom.*

The book of Esther, a story of Jewish persecution among Diaspora Jews, is a wonderful account of how a beautiful young woman saves her people from extermination. Esther, a beautiful virgin, was Jewish, a fact her uncle Mordecai had charged her not to reveal (Est 2:10). It came to pass that the king, Xerxes (Ahasuerus), so loved Esther from the moment he beheld her beauty, that he set a royal crown on her head and made her queen of Persia in place of the former queen, Vashti (Est 2:17). Sometime later she pleaded for the lives of her people, telling the king of the evil plans of Haman, the vizier to the king. The king had promised Esther anything, up to half of his kingdom, but she asked simply for the lives of the Jews (which would include herself) living in the Persian Empire. The king ordered the vizier hanged on the very gallows built by Haman to hang Mordecai. Esther's people were saved and her uncle (Mordecai) was made vizier in place of the executed Haman (cf. Est 7:1-10).

Harold Fisch (*Poetry with a Purpose*, p. 16) distinguishes between beauty which is "static" and that which is on the move. The movement of which Fisch refers takes a direction which "sheds grace" on the one in motion. The motion, therefore, has positive direction that is consistent with the ultimate strategy of God. The perfect example of beauty in motion is Isaiah 52:7, which I have mentioned earlier. But just in case, here it is again.

> How beautiful upon the
> mountains
> are the feet of the messenger
> who announces peace,
> who brings good news,
> who announces salvation,
> who says to Zion, "Your God
> reigns." (*NRSV*)

This type of beauty might be contrasted with that of the Shulammite (Song of Solomon), which is more static. She and her lover, says Fisch, "seem to circle around a still center." They end up where o betray Jesus, an event which depicts the exact opposite of the service of the woman. In the true biblical sense, the opposite is evil.

they began, with no direction and purpose, other than the physical enjoyment of the moment.

Proverbs 31:30 makes it clear that, "Charm is deceitful, and beauty is vain, but a woman who fears the Lord is to be praised." That is to say, by implication, that real beauty can be found in the one who fears the Lord. Otherwise, "all flesh is grass, and all its beauty is like the flower of the field, the grass withers and the flower fades" (Is 40:7-8).

Beauty and sexual imagery depict an allegory of love to illustrate God's fidelity to Israel, but the people do not respond as a devoted lover should. The word of the Lord came to Ezekiel concerning the failed love affair between the Lord and his people.

> You grew up and became tall and arrived at full womanhood; your breasts were formed, and your hair had grown; yet you were naked and bare.
>
> I passed by you again and looked on you; you were at the age for love. I spread the edge of my cloak over you, and covered your nakedness: I pledged myself to you, says the Lord God, and you became mine...
>
> I adorned you with ornaments: I put bracelets on your arms, a chain on your neck, a ring on your nose, earrings on your ears, and a beautiful crown on your head...
>
> You grew exceedingly beautiful, fit to be a queen...but you trusted in your beauty, and played the whore because of fame, and lavished your whorings on any passer by.
>
> (Ez 16:6-15, *NRSV*)

Harold Fisch remarks that beauty "is the thread on which salvation is strung." Then he makes the rather startling assessment that "if we follow that thread it takes us away from the beautiful into some other region" (*Poetry with a Purpose*, p. 17). Beauty, in the sense of Fisch's appraisal, plays a major role in attractiveness. That is to say we are drawn to physical beauty for the ultimate purpose of moving beyond it to a higher realm which transcends all outward appearances. The beauty of a cathedral leads us to holiness, or the beautiful music of praise brings us to a greater appreciation of the one to whom the praise is due. Even in the case of carnal love, the beauty which originally attracts two lovers is eclipsed by the spiritual oneness of sacred marriage. Without the deeper love between two individuals that transcends physical attractiveness, there is no love at all. It can be said that beauty is something illusive, a goal that humankind chases after, only to discover no lasting satisfaction in it. Then it becomes a matter of direction, as Harold Fisch points out. In what direction does that discovery lead us? Do we continue to circle around a stagnant center, going nowhere, or do we move beyond what we originally sought after in physical beauty to that which is ultimately beautiful?

Ezekiel had a problem in that his oracles were almost too beautiful. His audiences came to hear him because his words and their delivery were like a musical entertainment, yet they paid no attention to the content of what he said. They saw the beautiful covering, the performance of an artist, but failed to recognize the flesh and bones of what lay beneath. They listened but did not hear. Ezekiel's audience is typical of the "spiritually deaf" person we spoke of in Chapter Five.

As for Ezekiel, he rejects his role as a minstrel—one who "sings flute songs," but as he says this he cannot resist his fascination with words, asserts Fisch (p. 44). After remarking plaintively, "Ah Lord God! they are saying of me, 'is he not a maker of allegories?'" After this he proceeds to the beautiful oracle about the sharpened sword:

> A sword, a sword is sharpened,
>   it is also polished;
> It is sharpened for slaughter,
>   honed to flash like lightning!
> How can we make merry?
>   You have despised the rod,
>   and all discipline.
> The sword is given to be
>   polished,
> to be placed in the slayer's hand.
>   (Ez 21:9-11).

Returning to the Song of Solomon, the beauty of the Shulammite is static. It has no movement to it, no saving grace is shed by the woman in her beauty. If taken in a allegorical sense, however, the poetics take on a whole new meaning. Indeed, it is difficult to read these verses as a purely human construction, since the depth of love expressed is far beyond the purely human emotion. The love here is not in any sense a lighthearted game, as the lover says:

> Set me as a seal upon your heart,
>   as a seal upon your arm;
> for love is strong as death,
>   passion fierce as the grave.

CHAPTER ELEVEN

The Resurrected Body

> *When I look at your heaven, the work of your fingers, the moon and the stars that you have established; what are human beings that you are mindful of them, mortals that you care for them?* (Ps 8:3-4, *NRSV*)

In Chapter One, the changing perspective on the human body was traced from the period of ancient Israel to New Testament times. There seems little doubt that the views of the ancient Israelite regarding the human body were profoundly different from, say, the perspective of the apostle Paul. In the days of the patriarchs down to the period of Israel's monarchy, the "body," as it has come to be understood among first century Christians, did not exist. As has been said repeatedly, there is no separate word for "body" in the ancient Hebrew language.

Beginning with the eighth century prophets, there was a rising tide of religious consciousness, a phenomenon which increased in fervor just before and during the Babylonian exile. As the relationship of Yahweh to his people came to be seen in a different light, there occurred a growing appreciation for the dignity of the individual person. Jeremiah and Ezekiel, speaking for the Lord, promised an unprecedented event to come into the lives of the people. God was going to write his laws on their hearts and make a new covenant with them (Jer 31:33). Stony hearts would be removed to be replaced by a new heart and a new spirit (Ez 36:26). Jeremiah and Ezekiel thus prepared the way for a totally new outlook of the people, as regards their relationship to God and toward individual self-understanding. To be sure, the words of these great prophets anticipated the revolutionary new appreciation of God's love for all humanity, and what he was prepared to do for us in his Son,

Jesus the Christ.

Those developments bring us to the challenge of understanding what all this means in the modern age. What does it mean to be the "the temple of the Holy Spirit?" How should we understand Paul's idea of a "new creation?" Has the Christ-event really had an effect on the nature of the bodies we live in? "We are being transformed," writes the apostle Paul, and this transformation "comes from the Lord, the Spirit" (2 Cor 3:18). Should we now regard the faithful as some kind of matter-spirit hybrid? Paul says, "So if anyone is in Christ, there is a new creation, everything old has passed away; see, everything has become new. All this is from God, who reconciled us to himself through Christ..."(2 Cor 5:17-18, NRSV). The author of 1 John seems to be saying something similar, which I repeat here:

> Beloved, we are God's children now; what we will be has not yet been revealed. What we do know is this: when he is revealed, we will be like him, for we will see him as he is. And all who have this hope in him purify themselves, just as he is pure. (1 Jn 3:2-3, NRSV)

Guidelines for Speculaltion

I believe it is a mistake to dismiss out of hand speculative efforts concerning life after death. Often, the faithful Christian tends to shrug off such thoughts because they represent a central mystery that will be understood only at the time it is experienced first-hand. Non-believers, even in one of their kindest moments, would regard speculation on an afterlife as meaningless drivel. I am undaunted by such negative comments, however, because I am convinced in my own mind, that contemplation on the nature of the resurrected body is important. I believe it is a healthy exercise of our God-given intellect, and can be spiritually beneficial as well. Here I am not speaking as a expert on the subject, which I am not. Rather I am expressing a point of view that I hold dear to me.

Given this is not a particularly scholarly discussion, I believe, nevertheless, it is important that some guidelines for speculating on the resurrected body be observed. Otherwise, we might as well be writing in a science fiction or fantasy genre:

1. Conclusions must have some Scriptural basis. Otherwise, it might be tempting to allow the imagination to meander

unchecked as to the final disposition of the resurrected existence.

2. The speculation should involve a vast improvement in the earthly bodies we now inhabit. There would be little value, and certainly no hopeful anticipation, in a future existence of pain and suffering, war and strife. If these were to continue into the next life, most of us would opt for nothingness.

3. The nature of our existence, including the chemistry and physics of the present universe, must be the starting point for our speculative thoughts. Otherwise, we become entrapped in a fantasy land without rules or sense of direction.

4. Reason and common sense should prevail in the discussion, even though we are contemplating a profound mystery. Unbridled speculation is not speculation at all. Reason does not preclude faith, of course, but neither should we abandon the intellect.

I realize there is some overlap in the four basic guidelines I have offered above. Others might argue with me on this or that point, or add to those already made. I believe, however, the guidelines suggested are a good beginning.

Human Anthropology and the Image of God

In the study of man, *qua man*, (humankind as humankind), we inevitably encounter questions of an ultimate nature, particularly in regard to the question, *where does man the animal leave off and man the imago Dei begin*? Are we indeed simply a bag of water, biochemicals, etc., organized to speak and reflect on our own origins, or are we something more?

The proper study of humankind, as has been asserted, is humankind, and in a certain sense that is *all* human beings are capable of knowing. God, for example, is the ineffable *Other* of whom we can know nothing beyond the symbolic representations that come to us within the human context. To study God, in this sense, is to study ourselves, because all we can ever hope to know about God is that which comes out of our limited, human understanding. The very fact that we are capable of this kind of pursuit tells us something of ourselves. The conclusions we reach, varied as they may be, tell us even more about the nature of our humanity. We learn, among other things, that we are truly *Homo religiosus*, in a psychological sense at the very least.

This view of humankind is not limited only to esoteric theological conclusions. It applies as well to our assessment of all creation, from the simplest to the most complex structures. Take sodium chloride (NaCl), the very humble and chemically simple molecule of *table salt*, so familiar in everyday experience. What we know about this substance, whether from the point of view of its chemical analysis, taste, or appearance, all boils down to a *phenomenon,* as Immanuel Kant would say. It is how NaCl presents itself to human analysis, rather than any absolute understanding of the reality-in-itself as this might apply to salt (*noumenon*). Even in the examination of the humble table salt, therefore, in the final analysis we are studying human perceptions as well.

Anthropology (Greek, *anthropos*, meaning man), therefore, is the study of humankind, but it comprises much more than anatomy, biochemistry, psychology, history, social standards, and so on. It also includes all of our abstractions and conclusions about things human and non-human, natural and supernatural. It can therefore be said that when we study theology, we are in a sense, studying ourselves. Or, expressed in other terms, the more we know about the divine milieu, the more we know about the dimensional limits of humankind. Turning that same proposition around, it might be said that the more we learn about creation the more we understand the Creator. Many would say there is a spiritual message in creation, if only we will read it. Priest-anthropologist Teilhard de Chardin (*The Phenomenon of Man*), asserts that creation has a privileged axis and a precise orientation, especially a creation which includes human beings as the pinnacle of that creation.

Anthropology seeks to answer many questions about humankind, such as: How does the human person integrate his feelings and drives; desires, intellect and judgment; sexuality, memory and will? Is one *part* more godly than another, more an image of the Divine? Is there a spiritual ground of existence present in all human beings, an essence that transcends our physical existence? There are also questions about human motivation and activity, what should be the ultimate goal of human individuals and how does time and history relate to this goal?

Truly, there are no objects of study that are not related to anthropological concerns, nothing in the universe that we can totally divorce from human scrutiny and activity. Indeed, from our perspective, everything that exists has an *anthropological twist*.

Faith Seeking Understanding.

It can be safely concluded that, at the very least, human beings are more that the *sum of their parts*. This is not saying a great deal, however, about humankind as the pinnacle of creation. We can say as much about any lower animal, or even of a human creation such as a radio or television. Functionally at least, all of these are more than the sums of their parts

And yet we humans, reflecting on the ultimate questions about the purpose of our very existence, sense that we are special among God's creatures, and that somehow we play a very important, if not indispensable role in the God's overall strategy for creation. This sense of uniqueness, of being special among God's creatures, comes not from any sense of biological superiority, but from what we are calling *faith*. Faith is the key to human transformation. It is *faith itself* that provides humankind with the opportunity for salvation, for *eternal life*.

Based on Scriptural studies, most theologians agree that *faith is a gift* from God. This important biblical concept has led to the theological corollary that *one cannot separate the gift from the Giver*. That is to say, the believer must come to the conclusion that the transformation of our bodies is one of *theopoiesis* (from Athanasius), or divinization. In simpler terms, God takes up residence, comes to live in our bodies, sends us the Holy Spirit because, as the apostle Paul says, we are temples of the Holy Spirit. Faith is therefore not simply a change in mental attitude. In effect, there has been a new *creation* (cf. 2 Cor 5:17).

Paul's insight of the new creation (2 Cor 5:17; Gal 6:15), manifest itself in the transformed person of faith, and will be realized in hope, for although it is as yet unseen, it is real. Speaking of this transformation through the Spirit, Paul writes, "we all, with unveiled face, beholding the glory of the Lord, are being changed into his likeness from one degree of glory to another—" (cf. 2 Cor 3:18). The Greek word the apostle uses to express the idea of "we are being changed" is metamorphoumetha. It is therefore a concept which represents the whole human person caught up in a transformation that divinizes, rather than consisting of a simple *change*. As Pope John Paul II explains this remarkable insight (*Crossing the Threshold of Hope*, 22), "the work of redemption is to elevate the work of creation to a new level. Creation is permeated with a redemptive sanctification, even a divinization." These statements by the Apostle and the Pope help us to better understand the enigmatic words of 1 John 3:2 (see above).

Pope John Paul II also writes in his book, that "Christ is the sacrament of the invisible God—a sacrament that indicates presence." An *invisible* God is another way of saying a *hidden* God, but we might ask ourselves, "why is God hidden? Why is God not more accessible to us? Why can't we dial 911 for God, or see and hear God speaking on a television news program? Why doesn't God *zap* evil doers with a bolt of lightening?" After Thomas had seen and then believed, the Lord said to him, "blessed are those who have not seen and yet believed" (Jn 20:29). God is *invisible*, therefore, for a definite purpose, and that is because a visible God would remove the necessity of faith, perhaps even destroy our free will. It is our faith and the free assent to God's gift of faith that makes us blessed, and we are blessed because our faith transforms us.

The Image of God (Imago Dei).

Early Christian preachers and theologians borrowed the phrase, *image of God*, from the Hebrew Bible when they began to talk about the human person. In contrast to the pagan mythologies that surrounded the early Israelites, the Hebrew creation stories teach that the human person is created deliberately by *One God*, the source of all created things, and that this human creature is to be a part of a universe made harmonious by divine intention and governance. Another important theme of the Hebrew Scriptures was the goodness of all things.

How different this biblical anthropology was from that of the Mesopotamian gods and other pagan myths which portrayed the creation of the world and human beings as a capricious act which resulted from conflict among various deities. The resulting pagan view of the world was one of inherent chaos and instability because of the continuing conflicts between the evil and arbitrary forces of the gods. By contrast, when Genesis speaks of humans as *ruling* other creatures, it establishes a relationship between God, who rules the universe, and humanity, who is given the responsibility of stewardship over created things. Humankind, therefore, exercises a god-like role not given to other creatures of the world.

Many early Christian thinkers, like Justin, Theophylus, Tatian the Syrian, and Clement, appropriated Hebrew anthropology to support the Christian faith against popular ideas which threatened it (Garascia, "Theological Anthropology," 110). The Christian response of the first three centuries to heresies such as gnosticism and Arianism expanded the Christian understanding of the person as *image of God*. Cyril of Alexander (c. 380-444) is known for his teaching of various ways in which humans can be envisaged as *images of God.* Human beings, according to Cyril, possess the faculty of reason, are capable of virtue and mercy, and have power over all things on earth (Garascia, 111). Thus we humans possess certain *derived* powers of creation, and in that sense we create great works of art, build bridges, houses, chairs, automobiles, radios and televisions. We are not creators in the same sense as God is *Creator*, but we certainly do create in a secondary sense.

Free choice, or the act to freely choose, to say yes to God, it seems to me, is a most important qualification for humankind as the *image of God.* However, according to Cyril, and for his contemporary, Augustine, it is not sheer freedom of choice which images God, but it is in the use of this freedom to choose virtue, or the good. God is totally free, and any freedom that humankind possesses must come to us in the sense that it images, albeit imperfectly, the freedom of God.

The final property of God the Creator, in which faith tells us we have a share in imaging, has to do with death, resurrection and eternal life. There is that something, call it soul, call it divine principle, that is incorruptible. It is that special aspect of the human essence or higher status which results from God's grace, the sharing of his divinity, that renders the faithful capable of life eternal. Otherwise we are, as the apostle Paul writes, mere earthen vessels (2 Cor 4:7). There is a great deal that has been written on this subject and it is not possible to examine all the various viewpoints, but there does seem to be a consistent pattern of belief regarding the nature of human immortality.

In her book entitled, *The Resurrection of the Body*, Caroline Bynum insists repeatedly that Christians have clung to the literal notion of resurrection despite attempts by theologians and philosophers to spiritualize the idea. There seem to be many and diverse cultural reasons for this viewpoint. In the fourteenth century, for example, not only were spiritualized interpretations firmly rejected, writes Bynum, the "soul itself was depicted as embodied." Indeed, there was enormous importance attached to proper burial "because of a need to preserve difference (including gender, social status, and personal experience) for all eternity" (Bynum, xviii, 9).

The gospel accounts allow for a range of interpretations as to the status of Jesus' resurrected body. They stress the material form of the body which ate broiled fish and commanded Thomas to "handle me" (Lk 24:39-43). By contrast the accounts also emphasize the radical transformation of the resurrected Christ, who passed through closed doors (Jn 20:19), was not recognized by his beloved disciples (Lk

24:16), and mysteriously vanished out of their sight after final recognition (Lk 24:30-31).

We have the apostle Paul to thank for the most detailed explanation of what the resurrected body will be like:

> But some one will ask, "How are the dead raised? With what kind of body do they come?" You foolish man! What you sow does not come to life unless it dies. And what you sow is not the body which is to be, but a bare kernel, perhaps of wheat or of some other grain. But God gives it a body as he has chosen, and to each kind of seed its own body. For not all flesh is alike, but there is one kind for men, another for animals, another for birds, and another for fish. There are celestial bodies and there are terrestrial bodies; but the glory of the celestial is one, and the glory of the terrestrial is another. There is one glory of the sun, and another glory of the moon, and another glory of the stars; for star differs from star in glory. So is it with the resurrection of the dead. What is sown is perishable, what is raised is imperishable. It is sown in dishonor, it is raised in glory. It is sown in weakness, it is raised in power. It is sown a physical body, it is raised a spiritual body. If there is a physical body, there is also a spiritual body. Thus it is written, "The first man Adam became a living being"; the last Adam became a lifegiving spirit. But it is not the spiritual which is first but the physical, and then the spiritual. The first man was from the earth, a man of dust; the second man is from heaven. As was the man of dust, so are those who are of the dust; and as is the man of heaven, so are those who are of heaven. Just as we have borne the image of the man of dust, we shall also bear the image of the man of heaven. I tell you this, brethren: flesh and blood cannot inherit the kingdom of God, nor does the perishable inherit the imperishable. Lo! I tell you a mystery. We shall not all sleep, but we shall all be changed, in a moment, in the twinkling of an eye, at the last trumpet. For the trumpet will sound, and the dead will be raised imperishable, and we shall be changed. For this perishable nature must put on the imperishable, and this mortal nature must put on immortality. When the perishable puts on the imperishable, and the mortal puts on immortality, then shall come to pass the saying that is written: "Death is swallowed up in victory. O death, where is thy victory? O death, where is thy sting?" The sting of death is sin, and the power of sin is the law. But thanks be to God,

who gives us the victory through our Lord Jesus Christ
(1 Cor 15:35-57).

Paul expresses the awesomeness of salvation in paradoxical terms, e.g., the power of God is revealed as being "made perfect in weakness"(2 Cor 12:9-10). Here, in terms of the resurrection, Paul speaks oxymoron fashion of the *spiritual body*, because "if there is a physical body, there is also a spiritual body," and "what is sown in weakness, it is raised in power" (1 Cor 15:43:44). These kinds of seemingly contradictory statements work only in cases where they point to something outside ordinary human experience. They are *anagogic* in nature because we must seek a resolution to the apparent contradiction not in the material universe but on the supernatural scale of possibility. In other words, these kinds of revelations can only be appreciated from the perspective of *faith.*

Something in this Pauline scenario seems to guarantee that the subject of resurrection is truly *us*, but Paul says, "flesh and blood cannot inherit the kingdom of God, nor does the perishable inherit the imperishable." How do we then reconcile this difficulty? If we are no longer *flesh and blood*, and we are *imperishable*, how can this resurrected *something* be truly us?

There is all sorts of resurrection imagery which skirts around this issue, such as the "flowering of a dry tree after winter, the donning of new clothes, the rebuilding of a temple, the hatching of an egg, the smelting of ore from clay, the reforging of a statue that has been melted down," and so on (cf. Bynum, 6).

All of these theological concerns stem from our very human approach to the problem. We too easily place human limitations on God, forgetting what is possible to God. If we take Paul at his word and translate 1 Cor 15 into modern parlance, we will be clothed (at the resurrection) in a *spiritual body*. Let us therefore not limit our conception of a "spiritual body" to something we ourselves have experienced. From our imagination, therefore, let us construct a *spiritual body* from *spiritual atoms, molecules, tissues and organs*. In other words, this resurrected body is (would be) composed of the same *elements* as was the physical body, with one important difference. In the "spiritual body" we will have complete control of the component parts (*spiritual atoms, molecules,* etc.), such that we may experience pain if that is our desire, or we can place ourselves in the middle of a raging inferno and never feel its effects on our imperishable bodies. We can visit other planets (if such still exist), travel at infinite speeds to distant galaxies and other universes, walk through walls, accompany the angels to deliver God's messages, visit historical times or transcend history, and, of course, bask in the glory of God's infinite beauty.

In our present condition on earth, our nervous systems control our musculoskeletal system, and inform us of our surroundings. This allows us to walk around, speak and hear, see and feel. By contrast, the spiritual body will be able to control the very "atoms" and "molecules" that make up the body. Thus, we can make ourselves into a solid, resembling our former life, or cause our atoms to spread out to form a gaseous substance. We can take food and have a functioning digestive system, or we can choose not to do so. Indeed, the resurrected body may opt to have all the tissues and organs function just as they did before physical death. Perhaps we may need to do this for awhile, just so we can adapt to our new existence in our new home of spiritual dimensions.

If all this sounds *far-fetched*, consider the words of St. Paul, for "what no eye has seen, nor ear heard, nor the heart of man conceived, [that is] what God has prepared for those who love him" (1 Cor 2:9; cf. Is 64:3). What Paul is saying here, it seems to me, is that no matter how wonderful our dreams of eternal life as a resurrected *body* might be, the reality of this existence will far exceed anything our limited intellects and emotions can bring to consciousness. It is these very words which gave the early martyrs of the church the courage to die rather than renounce Jesus Christ as their Lord (see Chapter Five for more on this subject).

Great misunderstandings of the power of God in resurrecting the body cropped up in the middle ages in the form of some very bizarre practices. It crept into the folklore that God somehow needed to reassemble the resurrected body from the original parts, putrefied or not. This mentality was reflected in late medieval curiosities such as entrail caskets, finger reliquaries (cf. Bynum, 11).

Preachers and theologians may pride themselves on avoiding reference to any *body-soul* dualism, but pious talk at funerals usually relate to the departed person surviving as a vague, benign spirit or as a thought in the memories of others. This practice is understandable, believes Bynum, since it is clear that the resurrection of the body is a doctrine that causes "acute embarrassment," even in mainstream Christianity. Yet, it appears certain that analysis of current philosophical discourse and contemporary culture suggests "that Americans, like medieval poets and theologians, consider any survival that really counts to entail survival of body" (Bynum 14, 15).

All the above images, some of them quite abhorrent, bring us to a consideration of death itself. Again, we may have a great many misconceptions concerning death, simply because we must, out of necessity, guess at what is happening. I would suggest that we examine death from a different perspective, from the point of view of what the dying person might experience. What is the transition like from physical body to a resurrected one? Of course, it is not possible to know, but that doesn't prevent us from imagining how it might be.

The Dying Process.

The most difficult adjustment in our new existence might be to live in the absence of time. I doubt there will be clocks in the afterlife, since they would be meaningless. Perhaps adjustments such as this would involve some pain in the beginning. For example, I cannot even write about a timeless existence without using words that relate to time. But then, it may be that we will not be suddenly immersed in timelessness. There may be a transition.

It is natural to think of dying as an event that occurs suddenly, within seconds or minutes, especially in cases of severe trauma. But how do we know that to be true of the person experiencing death. Perhaps time is not measured by the same parameters. What seems to the living as seconds and minutes after the heart stops beating and brain death occurs may represent an exploded period to the dying person. I do not mean to imply that the suffering is prolonged, I am speaking of that interval following loss of consciousness, that is, the transition period from life to death.

It may even be that we die in much the same way as we gestate in our mother's womb. Perhaps there is even a *spiritual umbilicus* that gradually introduces us to our future home in the afterlife. The more I think about it the more sense it makes to have the comfort of a spiritual nurturing from God (the umbilicus) during the gestation of our spiritual bodies. Otherwise, the shock of going from a time ordered existence to a timeless dimension would seem an almost impossible hurtle.

In this vein, perhaps the Alzheimer's patient already has "one foot in heaven." If not, where does that lost intelligence, his or her former self, reside? In severe damage to the brain, from disease or accident, how is the ego, the consciousness of who we are, maintained? Where is it kept until it is instilled into a "spiritual body?" I suggest that the intelligence of a person so disposed has already entered the spiritual dimension, at least partially so, but the umbilical to the material world has not yet been severed. What we see in the realm of the living defines for us the moment of death, whereas the person dying may be already experiencing a spiritual existence beyond death.

But you say, this wonderful transformation does not occur until the end of the world, when all will be changed as "the last trumpet sounds." To this I remind you that we are imposing a time limitation on death and the afterlife. From the last stick of the needle I remember before bypass surgery, until I awakened afterwards, there was absolutely no sense of the passage of time. It was as though less than a split second occurred, if that much, during the entire eight hours of my surgery and reawakening in intensive care. Now, there is no better anesthesia than death, at least as far as we know. In this sense, when we die, after the "umbilicus" has been cut, we may experience no passage of time before reawakening in a resurrected body.

One thing for sure, when we die we leave behind a life of sin and suffering. As I said, there is no better anesthesia than death. But why did we have to go through all those terrible experiences that most of us, at some time, must suffer through? What was the purpose of it all, and what does it mean to us in our new existence after death? Some would argue the point, but I believe the suffering of our human bodies is part and parcel of salvation. As mysterious as it may seem, it is an integral part of God's strategy for us. Otherwise, what other answer do we have for those who ask the question, "why do we have to suffer?"

Sin, Suffering and Immortality.

Some might say that sin and suffering (particularly suffering) are a kind of prerequisite for life after death. Certainly, most of us presume there is some purpose to life on earth. It suffering does not, in some manner, prepare us for a future life in God's kingdom, why does it happen?

There are various answers to this and other questions as they relate to the ultimate goal of living happily in our transformed existence, in our "spiritual bodies."

To the Buddhist, suffering is produced by clinging, by the desire for possessions in the form of wealth, health, life or even selfhood itself. Christians differ from this perception at two key points. First, for Christians, evil is a historical reality which every individual faces, and it causes suffering even though the inner disposition of the person may, to some extent, ameliorate the pain. Secondly, the propensity to do evil may be overcome only by the saving power of God, and not, as the Buddhist believe, by learning a discipline. To Christians, the cross symbolizes the destructive and ultimate death-dealing nature of historical existence. Whatever physical details the eternal life may involve (see above), from the very beginning Christians understood resurrection as a reversal of the twin evils of suffering and death.

Sin is the rupture of a relationship between the person, God, and the community. The sacrament of penance was introduced in the early church for the forgiveness of sins, especially among those who had denied Christ under the threat of martyrdom or severe persecution. Penitential practice began as a means of ensuring that penitents were truly sorry for their sins and would make adequate reparation, so as not to take lightly their sins or to squander the forgiveness of God. Joseph Martos (*Doors to the Sacred*, 321), says this was an important move toward modifying the church's rigid rule about only one reconciliation for Christians.

In the decades that followed, however, the general direction of ecclesiastical penitence was toward greater strictness and legalism. As a result, sin—which had earlier been viewed as a break in the relationship of love and trust between members of the community and God—was increasingly conceived of in legal terms, as a breaking of a divine law or the violation of an ecclesiastical law. In contrast to this legalistic view, Garascia (115) writes that the thinking of some of the patristic writers "is surprisingly modern." Irenaeas of Lyon, for example, argued that to be made in the *image of God* is to be created *immature*. Some falling into disobedience is to be expected and can even be educational. The direction the church took, however, was to favor the anthropology of Augustine who, as one of great experience on such matters, wrote a great deal about sin. It is not my purpose, however, to debate the various aspects of sin, especially as they might relate to the idea of *original sin*.

The most positive and uplifting view of sin, and one which I favor very much, however, comes from the pen of the fourteenth century English mystic, Julian of Norwich. Her views are positively uplifting because they make a connection between sin, and the suffering that comes with it (as do some modern writers, e.g., Christaan Beker in his book on *Suffering and Hope*).

In her most telling vision (showing), Julian sees humanity as the "servant of the Lord." The Lord sits in state, while the servant stands before him, eager to do his bidding. When the Lord sends him off on a task, the servant "dashes off, and runs at great speed, loving to do his Lord's will." But the servant immediately comes to grief. "He falls in to a dell and is greatly injured; and then he groans and moans and tosses about and writhes, but he cannot rise or help himself in any way." To make matters even worse it seems that the fallen servant cannot be comforted. The Lord is not angry with him. Rather he is filled with love and compassion, but the servant was unable to turn and face the Lord. Instead of drawing hope and consolation from his Lord, the servant is blinded by his own misery and distress. The pain of his body and his mind is so intense that not only could he not rise, he could hardly even remember his love for the Lord. He was completely isolated in his misery. Julian says of her vision, "I looked around and searched, and far and near, high and low, I saw no help for him."

Julian was unable to find the servant of her vision in any way blameworthy. Having noted this, she saw that the Lord himself imputed no blame to his servant, and was not in the least cross with him. Rather, the Lord planned a reward for the servant after he was rescued.

> Then this courteous Lord said this: See my beloved servant, what harm and injuries he has had and accepted in my service for my love, yes, and for his good will. Is it not reasonable that I should reward him for his fright and his fear, his hurt and his injuries and all his woe? And furthermore, is it not proper for me to give him a gift, better for him and more honorable than his own health could have been? Otherwise, it seems to me that I should be ungracious.
>
> (Jantzen, *Julian of Norwich*, 193)

So why does God allow suffering? It would be a perverse parent who allowed a child to undergo excruciating physical and mental pain, justifying that course with the promise that he/she will afterwards reward the child. Therefore, the only way it makes any sense, says Julian, is that the suffering must be intrinsic to the reward. Even sin, "to the degree in which the sin may have been painful and sorrowful to the soul on earth," as Julian is bold enough to say, "*will be rewarded*" (my italics).

Julian had to struggle with the message of her vision in contrast with the view of the church. Why is it, she asks, that the church must assign blame while God does not? She concludes that there is a sense in which we are not blameworthy before God, but in another, the church can best help us toward integration of the negative aspects of our lives (i.e., sin) by holding us responsible and attaching blame. But in terms of God's point of view:

> Our good Lord Jesus [has] taken upon him all our blame; and therefore our Father may not, does not wish to assign more blame to us than to his own beloved Son Jesus Christ. So he was the servant before he came on earth standing ready in purpose before the Father until the time when he would send him to do the glorious deed by which mankind was brought back to heaven.

Jantzen explains that the blame which Julian is speaking of when she says that the Lord Jesus has "taken upon him all our blame" is not the blame of God upon us. Just as God does not blame the Son, who is totally loved in his sight, no more does he blame us, who are equally loved, despite our wretchedness. The blame, rather, is a part of our own confusion, the blame we have brought upon ourselves. It is "the unproductive sense of guilt and worthlessness which make us feel that we are utterly unlovable" (Jantzen, 198).

When we sin, with all the accompanying suffering this implies, the sin is against ourselves rather than against God, since God attaches no blame in his love for us. But in our own brokenness we are unable to forgive ourselves. It is against this *self-sin* and *self-brokenness*, therefore, that Christ died for, rather than the appeasement of an angry God the Father.

Thus, Julian's theology is fundamentally different from that of Anselm (1033-1109), the first Bishop of Canterbury, in his most famous piece of writing, *Cur Deus Homo* (Why God Became Man). In Julian's subsequent reflection upon her vision, there is no idea of *penal substitution* in the Christ-event as it is in Anselm's view. It is rather our blame and self-loathing that Jesus has taken upon himself, coming to demonstrate as a human being his endless love of the Father for us, so that we may find dignity and worth and integration in that love.

There can be no better example of sin and suffering than found in the Bible itself. God's strategy uses all of his people, good and bad, in a grand saving gesture. Human frailty, sin and suffering, demonstrate repeatedly that our bodies are central to God's plan for us. The ambition of this small work is to provide evidence for this simple truth. If this conclusion is an accurate one, it seems a mistake to wait for God to "wipe away every tear" (Rv 7:17; 21:4). Christiaan Beker (*Suffering and Hope*, p. 30) warns we should be critical of cheap hope. Like

Bonhoeffer's rebuff of cheap grace (cf. *The Cost of Discipleship*), Beker says we should be diligently searching for "authentic Christian hope." It is not enough to simply endure the trials and tribulations of life on earth. Indeed, I believe it is incumbent on all who lay claim to faith in a loving, compassionate God, to assist in the saving strategy as befits individual talents and abilities. And since our bodies are the focal center to God's plan for us, we should make every effort to resonate with that ultimate goal, and in the process never lose hope.

Suffering must be tied to hope, for once the two are divorced from one another, suffering becomes naked and meaningless. Christiaan Beker (p. 37) points out that there is general and widespread consensus in Scripture, "that the experiences of suffering and hope cannot be sealed off and divorced from each other." They are, rather, "interdependent realities." In other words, Scripture reminds us that hope cannot be built on a foundation that denies the reality of suffering. Likewise, "suffering," says Beker, "is not to be 'suffered' without hope."

CHAPTER 12

A Partnership of Science and Religion

> *Suddenly, while he was traveling to Damascus and just before he reached the city, there came a light from heaven all around him. He fell to the ground, and then he heard a voice saying, "Saul, Saul, why are you persecuting me?" "Who are you, Lord?" he asked, and the voice answered, "I am Jesus, and you are persecuting me. Get up now and go into the city, and you will be told what you have to do." The men traveling with Saul stood there speechless, for though they heard the voice they could see no one. Saul got up from the ground, but even with his eyes wide open he could see nothing at all, and they had to lead him into Damascus by the hand.* (Acts 9:3-8).

A physician friend of mine once informed me of his belief that what the apostle Paul (Saul) had experienced (see above quote from the Acts of the Apostles) was a *temporal lobe seizure*. He was proposing a possible medical explanation of Paul's conversion experience, including not only the described fall to the ground and the blindness, but by inference, the "voice from heaven" as well. I got the impression at the time, although I did not pursue the point, that this "medical explanation," at least to my friend's way of thinking, negated the reality of Paul's conversion experience.

Reflecting on my friend's words over the years has led to a maturity in my thinking about the relationship of science and religion. In my book, *Science and Religion: Expelling the Demons from the Marriage Bed* (Factor Press, 2000), I discuss the various aspects of this relationship. It is too easy, perhaps, to think of science and religion as two separate camps, two isolated means of viewing reality. Nothing, of course, could be further from the truth. Indeed, most of us believe there is but one, all encompassing, reality. We perceive the totality of this one reality only dimly, but in our limited human capacity, we believe in it nevertheless.

God Working Through God's Laws

When my friend first made his remark concerning Paul's conversion experience, some thirty-five years ago, I may have been tempted to argue against his conclusion as to whether the Apostle actually experienced a seizure. In other words, there was a time in my life that I might have perceived the medical explanation of Paul's conversion experience, valid or not, as a threat to my Christian faith. I now see the whole incident as reflecting a degree of immaturity on my part, and as a great lesson in what is wrong with our present day perception of the proper relationship between science and religion.

Does it really matter, in fact, whether the Apostle had a seizure or not at the time of his conversion? Can we limit God to working through humans in certain, preconceived ways? Indeed, is it not possible for God to work in and through his human subjects by naturally occurring, physical means, or is it necessary for God to manifest some obviously miraculous event, such as light and voices from heaven? Obviously, we will never know if Paul actually experienced a seizure, nor does it make any difference in the final analysis.

Let us suppose that we could go back in time, and place upon Paul's head (unnoticed by the Apostle himself) some kind of minuscule and supersensitive encephalographic device, and then make a record of this hypothetical seizure. Would it make any difference as to our conclusion regarding whether or not Paul also experienced a revelation from God at that particular moment? The immediate reaction of a great many would consist of something like: "Paul just thought he had a revelation from God, but the fact of the seizure explains everything. Thus," they would continue, some with gleeful smugness, "the most profound theology in all the New Testament began from a simple cerebrocortical event." Historians of medicine would undoubtedly write volumes about the subject of how the Apostle's unfortunate medical condition changed the entire course of Western history.

One can also pose similar questions and objections about other New Testament events. Were the unclean spirits (demons) that Jesus casts out (Mk 1:23-26; 5:13) really some kind of supernatural possessions, or just an ordinary illness, or perhaps nothing at all? Were the "lepers" that Jesus made clean really lepers? Was the leprosy of first century Judea the same as the Hansen's disease of the present day? Did Jesus actually cure anybody of any real (physical) illness? One could ask these kinds of questions *ad nauseam* and many do, because they are the "bread and butter" questions of historical criticism. Biblical scholars ask these kinds of questions not because they are doubters (although some may be), but because they desire a better understanding of the historical times in which Jesus lived. Others seem to ask these questions in order to discredit the Christian faith, pure and simple.

A Unity from the Beginning.

Much as some might prefer for science and religion to be cordoned off from one another, representing separate and unrelated realities as it were, there are no grounds for such wishful thinking. Science and religion have been intertwined and often at odds with each other since the beginnings of civilization. These two great endeavors of humankind have been marriage partners and bedfellows since the dawn of history, with all the conflicts and mutual attraction the metaphor implies. As in any ideal marriage the two partners have grown and matured as a result of their encounters with each other. These growth spurts have occurred not always in spite of, but mostly because of, the conflicts. Both gathered their forces within a milieu of fear, violence, and arrogance.

Demons and Gods.

Diseases were demons, the unclean spirits of ancient civilizations. The gods were tricky, arbitrary and powerful. In the beginning the priest or shaman wielded the power, convincing the masses that he had the inside track to the special deity, or that he alone possessed the secret herbs that would expel the demons. In those days neither science nor religion held sway, one over the other, because there was really little difference between them, certainly none that was well defined. The Babylonians had their demons, herbs, and incantations, but they also possessed a burgeoning enterprise in astrology. This progenitor of modern astronomy held the key to many problems, including such diverse endeavors as healing the sick, counseling the king, or simply winning a battle. Of course, where there is astronomy there must also be mathematics, crude or refined. Indeed, all the sciences, if one can call them by such a modern term, were the bedfellows of religion.

One cannot say of these strange bedfellows that one was bad

and the other was good, or that one was right and the other was wrong, or even that one dealt with the possible and the other with the impossible. They were so interlinked, so interdependent in those ancient days, that one could not exist without the other. One could even say that it was the ideal situation, science and religion cooperating with each other in mutual respect and dependency. And in many ways it is a model for us in our day, despite the inadequacies and superstitions that governed the ancient system. It is a model, an ideal union one might say, because it consisted of a natural confluence of diverse interests, a marriage made in heaven, so to speak. So where do we stand now? What do science and religion have in common in modern times? Are there any hopeful signs for the future?

Mystical Beckonings From the Unknown

In humankind's ceaseless effort to understand itself, there is the ever present enigma of creation itself. What is it that surrounds us, this mixture of structure and amorphous stuff. We eat and breath it, walk and even float upon it. We wonder at its abundance and contemplate its beauty. In nature, its power for destruction is frightening in its majesty, while its ability to sustain us is awesome. Indeed, our curiosity concerning the material world in which we live has given rise to efforts to comprehend its awesomeness and majesty, and in the process we have developed a discipline we now define as *science*. The question is, how far can science go, how deep can it probe, in fathoming the mysteries of creation? Is there a limit to the depths we attempt to plumb. Is there sacred ground upon which we dare not tread in our quest for more and greater understanding?

The Limits of Knowledge.

There are some who are greatly disturbed by a scientific world that accepts no limits. The fear is that science will one day remove the veil from our most cherished mysteries, attacking the very core of our religious belief. One need only examine recent works in the fields of astronomy and astrophysics, where the creation of the universe and the "decrees of God" are mentioned in the same breath, to glimpse the origin of those fears. (cf. Stephen Hawking's, *A Brief History of Time* Bantam, p. 122). Chapter titles such as "Mathematicians and Mystics," or "Before Creation," (*Michio Kaku's Hyperspace*), are revealing enough to make the ordinary reader marvel at the audacity of the author. One might ask, "is nothing sacred?"

Beginning with the eighteenth century Enlightenment the confidence of scientists has frequently been expressed in terms of boundless exuberance for a glorious future in which there are no mysteries, merely unknowns yet to be discovered. Nowadays, whenever limitations are suggested for scientific inquiry, the idea is usually rejected by the scientific community as a whole. John Horgan, a longtime staff writer for *Scientific American,* has published an intriguing book, entitled, The End of Science: Facing the *Limits of Knowledge in the Twilight of the Scientific Age* (Helix, 1996). This work is of particular interest here, since it highlights a crucial argument concerning scientific limitations.

In a David Gergen interview on the Jim Learer news show (PBS), Horgan expanded on some of the ideas in his book. We will never understand what human consciousness really is, he explained, or what happened before the big bang (the primal explosion which gave rise to the universe). Horgan insisted that we have reached the edge of an abyss in what is humanly possible to gather and scrutinize as hard scientific data. Indeed, scientists are asking questions now that prompt answers in theological terms, and that is why there is so much "God-talk" in their books. The problem here is that scientists (as scientists) are not qualified to answer theological questions, although they attempt to do so on a regular basis. Be that as it may, I find Horgan's book encouraging, because it opens the door to a general discussion of a dimension beyond the physical universe. The unfortunate aspect of most books in this genre is that the scientific data are too sophisticated for the average person to grasp. The "average person," therefore, may continue to live under the false impression that science can solve most if not all his/her problems, perhaps even theological ones. This mentality, living as we do in the world of "on" and "off" switches, does not encourage one to conclude that the god of science is profoundly limited.

Given that absolute truths are not the proper object of physico-chemical experimentation, science cannot venture beyond the limits of its methods, regardless of how rigorous or sophisticated the discipline might be. Certainly it is proper, even necessary, that scientists endeavor to stretch disciplinary boundaries, and expand the horizons of science. But when science reaches its limits, beyond which there is no possibility of further progress, this is a discovery in itself. It is here that the limitless *beckons* to the investigator across the chasm between the types of wisdom of which the apostle Paul speaks so eloquently (1 Cor 1:22-25). What can be troubling to some of us is the arrogance of those scientists who never admit to any such limits. Their wisdom is the only wisdom, they believe, and it recognizes no boundaries. That is why it is refreshing to learn that some scientists accept the reality of limits, and even take the trouble to define certain of them.

In all scientific experiments it is necessary to somehow get outside the specimen being studied. In order to examine a cell, for example, the researcher has to remove it from the animal (or plant), place it under a microscope, perhaps stain it with a suitable dye, then describe the structures observed. The biochemist usually grinds up the cell and looks for the presence of proteins, fats, amino acids, sugars, hormones, etc. Thus it is necessary to destroy the living unit in order to learn all that is possible to know about a particular specimen. In preparing the cell, so that it is no longer living, scientists lose the ability to totally understand the specimen. This method of cell and tissue analysis, therefore, is a destructive procedure, and always will be so to some extent. I have studied cellular structure through electron microscopes for more than thirty years, and I know from experience that the smaller the object being examined, the more destructive the analytical procedure becomes.

The closer we approach the very basis of life, or the very basis of material existence, even down to atoms and subatomic particles, the more the principle of specimen destruction holds true. We must somehow get outside the object of the experiment in order to understand it from the limited viewpoint of science. Carrying this principle a little further, to reach a thorough appreciation of matter and energy (if such were ever possible), the scientist would somehow have to become nonmaterial to do so. In other words, to completely understand the fundamental forces and basic nature of material existence, a spiritual being would be required to carry out the experiment. This is indeed a limitation of the scientific method, one which we don't hear much about.

Is evolution a process of blind chance, or is there an intelligence behind it? What happened before the big bang, the commonly accepted beginning of time and the material universe? Why is there something rather than nothing? Is there truly a spiritual dimension to existence of which the prophets and mystics speak? Why do not more investigators recognize that it is a worthy objective to define the limits of empirical science? Why is there not more said or written about those mysteries of the universe that will never be solved by scientific methods? Let other scientists try to "shoot down" these theories, argue fundamental points and reach tentative conclusions about the interface between physics and metaphysics, of what is knowable and unknowable from scientific instrumentation.

We have come a long way since the eighteenth century, when there was such confidence in science that it seemed to the unsophisticated empiricist that there were no mysteries, only things yet to be discovered. The hangover from that time, however, is the absolute disdain of many scientists for anything religious. It is this kind of scientist, in his/her arrogance, that sees no need for a personal God, or any of the "trappings" of religion. He/she seems perfectly happy living in a godless world where science rules. When science rules, of course, the scientist becomes the ruler, if not the philosopher king.

I do not intend to paint scientists as so tied to their disciplines that they refuse to consider theological questions. Many believe, as in my case, in a loving, personal God, and some who do not, *wish they could.* But the scientist, like the average modern person, is full of the business of day-to-day activity. He/she may be so bound up in a personal timetable, in the functions of his/her body, that theological questions are simply pushed to the side. In this sense, the scientist is no different from the average non-scientist.

Pushing the Envelop of Science.

Learning about the natural world through the instruments of science has opened up new horizons for humankind to contemplate. As instruments become more sophisticated, sensitive, and precise, they nevertheless provide us with a clearer picture of the scientific barriers that cannot be crossed. I am not suggesting that humankind should give up on attempting to cross them, but in the end it is simply a matter of our incapacity to do so. Indeed, humanity must continue to press upon the boundaries of science in order to finally define the limits of experimentation and its corresponding lack of useful data. Of course, what appears to us a present day barrier may not apply to succeeding generations of scientists. On the other hand, it doesn't matter how expensive or how big the instrument science might contemplate in the future, there is only so much we can learn about the basic building blocks of matter. I agree with John Horgan that science can never tell us what preceded the big bang, or define the electro-chemical basis for the human ego, although I am sure there will be books written on these very subjects.

Using their sophisticated instruments, physicists will always try to penetrate the impenetrable, because the unknowables at the threshold between matter and spirit keep beckoning to them.

Humankind has an intense attraction for mystery, and always will have. We are a profoundly curious species, and this curiosity, it seems to me, must spring from mystical roots. Science can go only so far in its quest to break the code of mystery, but that doesn't mean that scientists will relinquish their appetite for the unknowable. Books like Frank Tipler's *The Physics of Immortality* (Doubleday, 1994) are evidence for what I am saying here. That is, that science is driven toward these unknowable objects by mystical forces that none of us, especially the scientist held in the grip of these forces, can fully comprehend.

In my earlier years I would have counted this Quixotic desire to conquer all barriers as an offensive and unpardonable attitude on the part of many scientists. I would not say that at the present time, because I now understand more fully the irresistible call to explain the inexplicable, to understand the impossible, to plumb the depths of that mysterious river that seems to divide the material world from the spiritual dimensions of reality. From a faith perspective, it makes sense that God places in human creatures that desire to know more, to go beyond the possible, to seek out mysteries, to reconcile insurmountable paradoxes, even to probe the "cloud of unknowing," the mind of God. This attraction for mystery, therefore, is not just God challenging us, it is God wanting to share his divine nature with us, yet requiring us to meet him part way by dint of willful effort.

The scientist is therefore drawn inexorably toward the mystical beckonings that lie on the other side of the scientific barriers imposed by nature. When all scientific instruments fail to penetrate this barrier, when theoretical physics has exhausted all of its options, there is no other way to grow but through the revelations that come from religious faith.

The non-scientist, always at the periphery of these ultimate arguments, but an observer nonetheless, will have profited from them as well. Indeed, humankind may be at the threshold of a new spiritual awakening in which science and religion are becoming less and less adversarial. Rather, these two great forces for truth may be viewed more as cosmic marriage partners instead of unremitting enemies. It is certainly a less than perfect and harmonious union, but a union nevertheless, and one that portends a mutual understanding never before possible.

BIBLIOGRAPHY

Anderson, Bernhard W., *Out of the Depths*. Philadelphia: Westminster Press, 1983.

Aristotle Dictionary, ed. by Thomas P. Kiernan. New York: Philosophical Library, 1962.

A Concordance to the Greek Testament. F. Moulton, A. S. Geden, eds., Edinburgh: T. & T. Clark, 1978.

Barrett, C. K., *A Commentary on the Epistle to the Romans*. New York: Harper and Row, 1957.

Beker, Christiaan, *Suffering and Hope*. Grand Rapids: William B. Eerdmans Pub. Co., 1994.

Birky, C. William, Jr. "Transmission genetics of mitochondria and chloroplasts." Ann, Rev. Genet, 12:471-512 (1978).

Blenkinsopp, Joseph, *The Pentateuch*. New York: Doubleday, 1992.

Bonhoeffer, Dietrich, *The Cost of Discipleship*. New York: The Macmillan Co., 1967.

Brennan, Robert Edward, O.P., *Thomistic Psychology*. New York: The MacMillan Co., 1954.

Bright, John, "Jeremiah," in *The Anchor Bible*. New York: Doubleday, 1986.

Brown, Raymond E., *The Community of the Beloved Disciple*. New York: Paulist Press, 1979.

_________________., *An Introduction to the New Testament*. New York: Doubleday, 1997.

Brueggemann, Walter, *Old Testament Theology*. Ed. Patrick D. Miller, Minneapolis: Fortress Press, 1992.

_________________, *A Social Reading of the Old Testament*. Ed. Patrick D. Miller, Minneapolis: Fortress Press, 1994.

Bynum, Caroline Walker, *The Resurrection of the Body*. New York: Columbia University Press, 1995.

Bultmann, Rudolf, *New Testament and Mythology*. Ed. and Trans. by Schubert M. Ogden, Philadelphia: Fortress Press, 1984.

Campbell, Joseph, *The Masks of God: Creative Mythology*. Penguin Books (1976).

Carmody, Denise Lardner. *Feminism & Christianity, a Two-Way Reflection.* Abington/Nashville (1982).

Childs, Brevard, *Old Testament Theology in a Canonical Context.* Philadelphia: Fortress Press, 1985.

______________, *Biblical Theology of the Old and New Testaments.* Minneapolis: Fortress Press, 1993.

Clift, Wallace B., *Jung and Christianity*. New York: Crossroad, 1982.

Cooke, Bernard J., *The Distancing of God.* Minneapolis: Fortress Press, 1990.

Corsini, Eugenio, *The Apocalypse, the Perennial Revelation of Jesus Christ.* Trans. and Ed. by Francis J. Moloney, S.D.B., Wilmington: Michael Glazier, Inc., 1983.

Cousar, Charles B., *A Theology of the Cross*. Minneapolis: Fortress Press, 1990.

Cranfield, C. E. B., *The Epistle to the Romans*. ICC, Vol. 1, Edinburgh: T. & T. Clark, 1975.

Christ, Carol P. and Judith Plaskow, eds, *Womanspirit Rising*. Harper & Row (1979).

Cunningham, Philip J., *Exploring Scripture*. New York: Paulist Press, 1992.

Currid, John E., "Why Did God Harden Pharaoh's Heart?" *Bible Review*. Dec., 47-51:1993.

Dudley, Donald, *The World of Tacitus*. Boston: Little, Brown and Co., 1968.

Donahue, John R., *The Gospel in Parable*. Fortress Press, 1988.

Dulles, Avery, *Apologetics and the Biblical Christ*. Paramus N.J.: Newman Press, 1971.

____________, *The Craft of Theology*. New York: Crossroad Publishing Co., 1992

Fisch, Harold, *Poetry with a Purpose*. Bloomington: Indiana U. Press, 1990.

Fitzmyer, Joseph, "Pauline Theology," in *The New Jerome Biblical Commentary*, ed. by Raymond Brown, Joseph Fitzmyer, Roland Murphy, New Jersey: Prentice Hall, 1990.

______________, *A Christological Catechism*. Revised ed., New York: Paulist Press, 1991.

Garascis, Mary M., "Theological Anthropology," in *Introduction to Theology*. ed. by Thomas P. Rausch, Colleveville: The Liturgical Press, 1993.
Garrison, Fielding H., *History of Medicine*. 4th ed. Philadelphia: W. B. Saunders Co., 1929.
Giblin, Charles H., "A Summary Look at Paul's Gospel: Romans, Chapters 1-8," in *A Companion to Paul*. ed. by Michael Taylor, Oxford: University Press, 1974.
Grant, Michael, *The Ancient Historians*. New York: Charles Scribner's Sons, 1970.
Grant, Robert, *Augustus to Constantine*. New York: Harper and Row, 1970.
____________, *Greek Apologists of the Second Century*. Philadelphia: The Westminster Press, 1988.
Harper's Bible Dictionary. ed. by Paul J. Achtemeier. San Francisco: Harper and Row, 1985.
Hawking, Stephen, *A Brief History of Time*. New York: Bantam Books, 1988
Holmes, Michael W., *The Apostolic Fathers*. 2nd Ed., Trans. by J. B. Lightfoot and J. R. Harmer, Grand Rapids: Baker Book House, 1989.
Horgan, John, *The End of Science: Facing the Limits of Knowledge in the Twilight of theScientific Age*. Helix, 1996.
Jantzen, Grace M., *Julian of Norwich*. New York: Paulist Press, 1988.
Jewett, Robert, *Paul's Anthropological Terms*. Leiden: E. J. Brill, 1971.
Jung, Carl G., *Man and His Symbols*. New York: Dell Publishing Group, 1964.
___________, *Psychology and Religion*. New Haven: Yale University Press, 1938.
Kaku, Michio, *Hyperspacey*. New York: Doubleday, 1994
Fiorenza, Elizabeth, "Feminist Spirituality, Christian Identity and Catholic Vision," in *Womanspirit Rising*. ed. by Carol P. Christ and Judith Plaskow, San Francisco: Harper and Row, 1979.
Kasemann, Ernst, *New Testament Questions of Today*. Trans. by W. J. Montague and Wilfred F. Bunge, Philadelphia: Fortress Press, 1969.
___________, *Commentary on Romans*. Trans. and ed. by Geoffrey W. Bromily, Grand Rapids: Eerdmans, 1980.

Keck, Leander E., *Paul and His Letters*. Philadelphia: Fortress Press, 1988.

Kelly, Geffry B., ed., *Karl Rahner*. Minneapoplis: Fortress Press, 1992

Kelly, J.N.D., *Early Christian Doctrine*. Revised ed., Harper San Francisco, 1978.

Kittel, Gerhard and Gerhard Friedrich, eds. *Theological Dictionary of the New Testament*. Trans. and abridged by Geoffrey W. Bromily. Grand Rapids: Eerdmans/Paternoster Press, 1988.

Klein, Gunter, "Paul's Purpose in Writing the Epistle to the Romans," in *The Romans Debate*. ed. by Karl P. Donfried, Peabody, Massachusetts: Hendrickson Publishers, 1991.

Kung, Hans, *Theology Into the Third Millenium*. Trans. by Peter Heinegg, New York: Doubleday, 1988.

Kuyper, Lester J., *The Hardness of the Heart According to Biblical Perspective*. Scottish J. Theol., 459-74, 1974.

Major, Ralph H., *A History of Medicine*, Vol. 1. Springfield: Charles C. Thomas, 1954.

Marcel, Gabriel, "On the Ontolgical Mystery." in *Philosophy in the Twentieth Century*, Vol. 3, ed. by William Barrett and Henry D. Aiken, New York: Random House, 1962.

Marrow, Stanley B., *Paul, His Letters and His Theology*. New York: Paulist Press, 1986.

Martos, Joseph, *Doors to the Sacred*. New York: Doubleday, 1982.

Matthews, Victor H. and Don C. Benjamin, *Old Testament Parallels*. New York: Paulist Press, 1991.

McKinzie, John L., "Aspects of Old Testament Thought," in *The New Jerome Biblical Commentary*, ed. by Raymond Brown, Joseph Fitzmyer, Roland Murphy, New Jersey: Prentice Hall, 1990.

Morton, Nelle. "The Dilemma of Celebration," in *Womanspirit Rising*. ed. by Carol P. Christ and Judith Plaskow, Harper & Row (1979).

Meier, John P. *Matthew*. Washington: Michael Glazier, Inc., 1980.

_____________, *A Marginal Jew*. New York: Doubleday, 1991.

Murray, John Courtney, S.J., *The Problem of God*. New Haven and London, Yale University Press, 1964.

Neyrey, Jerome H., *Paul in Other Words*. Louisville: Westminster/ John Knox Press, 1990.

Nicolas, Antonio de, *St. John of the Cross*. New York: Paragon House, 1989.

Patte, Daniel, *Paul's Faith and the Power of the Gospel: A Structural Introduction to the Pauline Letters*. Philadelphia: Fortress Press, 1983.

Perrin, Norman, *Jesus and the Language of the Kingdom*. Philadelphia: Fortress Press, 1976.

Perrine, Laurence, *Sound and Sense*. 7th ed. New York: Harcourt Brace Jovanovich Press, 1987.

Pieper, Josef, *Guide to Thomas Aquinas*. Trans. from German by Richard and Clara Winston, The New American Library, 1962.

von Rad, Gerhard, *The Message of the Prophets*. trans. by D.M.G. Stalker, New York: Harper & Row Press, 1967.

Rahner, Karl and Herbert Vorgimler, *Dictionary of Theology*. 2nd ed., New York: Crossroad, 1985.

Robinson, John A. T., *The Body: A Study in Pauline Theology*. Philadelphia: Westminster Press, 1952.

Rohr, Richard and Joseph Martos, *The Wild Man's Journey*. Cincinnati: St. Anthony Messenger Press, 1992.

Ryan, Francis, *The Body as Symbol*. Washington/Cleveland: Corpus Books, 1970.

Russel, D. S., *Divine Disclosure*. Minneapolis: Fortress Press, 1992.

Shannon, William H., *Seeking the Face of God*. New York: Crossroads, 1990.

__________________, *Silence on Fire*. New York: Crossroad, 1991.

Shackleford, John M., *The Biblical Heart: The Dynamic Union of Flesh and Spirit*. Factor Press, 1996.

_________________, *Science and Religion: Expelling the Demons from the Marriage Bed*. Factor Press, 2000.

Stendahl, K., *Paul Among Jews and Gentiles and Other Essays*. Philadelphia: Fortress Press, 1976.

Stuhlmacher, Peter, "The Purpose of Romans," in *The Romans Debate*. ed. by Karl P. Donfried, Peabody, Massachusetts: Hendrickson Publishers, 1991.

Swete, Henry Barclay, *An Introduction to the Old Testament in Greek*. Peabody Massachusetts: Hendrickson Publishers, 1914.

Teilhard de Chardin, Pierre, *The Phenomenon of Man*. New York: Harper and Row, 1965.

_____________, *The Divine Milieu*. New York: Harper and Row, 1960.

The Cloud of Unknowing. Intro. by Clifton Wolters, New York:

Penguin, 1961.

Turner, V. W., *The Forest of Symbols*. Ithaca NY: Cornell University Press, 1967.

Vine, W. E., Merril F. Unger, William White, Jr., *Complete Expository Dictionary*. Nashville: Thomas Nelson Pub., 1996.

Whiteley, D. E. H.,*The Theology of Paul*. Oxford: Basil Blackwell, 1964.

Wolff, Hans Walter, *Antropology of the Old Testament*. Philadelphia: Fortress Press, 1974.

Yarbro Collins, Adela, *Crisis and Catharsis: the Power of the Apocalypse*. Philadelphia: The Westminster Press, 1984.

Yarbro Collins, Adela, "The Apocalypse (Revelation)" in *The New Jerome Biblical Commentary*. Ed. by Raymond E. Brown, S.S., Joseph A. Fitzmyer, S.J., and Roland E. Murphy, O.Carm., New Jersey: Prentice Hall, 1990.